Handbook of Electrochemistry

Handbook of Electrochemistry

Edited by **Jina Redlin**

New York

Published by NY Research Press,
23 West, 55th Street, Suite 816,
New York, NY 10019, USA
www.nyresearchpress.com

Handbook of Electrochemistry
Edited by Jina Redlin

International Standard Book Number: 978-1-63238-237-5 (Hardback)

Printed in the United States of America.

Contents

Preface

Electrochemistry is the branch of chemistry which is concerned with the relations between electrical and chemical phenomena. Electrochemistry has witnessed significant advances in the last few decades. It has now acquired the position of a domain of interest for a diverse range of spheres varying from academicians concerned with studying thermodynamic properties of solutions to industrialists involved in utilization of electrolysis or manufacturing of batteries. It has now become evident that these apparently distinct subjects, alongside others, have an element of commonality among them and that they have grown towards each other, particularly as a result of ongoing research regarding the rates of electrochemical processes. These advances have resulted from a number of factors and indicate a potential of carrying out repeatable, dynamic experiments under an ever-increasing variety of conditions with reliable and sensitive instrumentation. This has enabled various researches of both fundamental and applied nature, to be carried out in this intriguing domain.

Significant researches are present in this book. Intensive efforts have been employed by authors to make this book an outstanding discourse. This book contains the enlightening chapters which have been written on the basis of significant researches done by the experts.

Finally, I would also like to thank all the members involved in this book for being a team and meeting all the deadlines for the submission of their respective works. I would also like to thank my friends and family for being supportive in my efforts.

Editor

Chapter 1

Membrane Electrochemistry: Electrochemical Processes in Bilayer Lipid Membrane

Mohammed Awad Ali Khalid

Additional information is available at the end of the chapter

1. Introduction

Since the first report of reconstituted lipid bilayers or bilayer lipid membranes (BLMs) in vitro, more than 30 years have elapsed. It is informative, in retrospect, to mention in this introduction the crucial role played by the science of interfaces in the reconstitution experiments. In living cells, the tremendous interfacial areas that exist between the membrane and its surroundings not only provide ample loci for carrying out activities vital to the living system, but afford a clue for our understanding. Physically, an interface is characterized most uniquely by its interfacial free energy, which is a result of the orientation of the constituent molecules [1]. An ultra-thin film such as a lipid bilayer is a system whose interior is influenced by the proximity of its interfaces. In a sense, an interface can be thought of as a structure so that it has no homogenous interior. However, the kind of ultra-thin films (e.g., BLMs) under discussion here are heterogeneous from their contacting phases. This and other interfacial properties of membrane can be understood to a large extent in terms of the laws of interface chemistry and physics that govern them, in particular electrochemistry [2]. In this connection it is worth noting that within the last few years, several monographs on various aspects of electrochemistry have been published [2-4], attesting the importance as well as research activities on the state of the science. Indeed, electrochemistry embraces the field ranging from the theoretical to experimental, surface science, spectroscopy, and to ultra-thin film-based applications. For instance, cyclic voltammetry, one of the most elegant and powerful techniques of electrochemistry, was applied to the BLM research in 1984 [5]. A planar BLM is a 5 nm thick lipid bilayer structure separating two aqueous solutions [6], which along with the spherical liposom, has been extensively used as an experimental model of biomembranes [7-13]. In fact, the current understanding of the structure and function of biological membrane can be traced to the investigations of experimental model membranes which have been developed as a direct consequence of the applications of classical principles of interfaces advanced by Langmuir,

Adam, Harkins, McBain, Hartley, and others [1, 6, 7, 14]. In the last few years there have been a number of reports on self-assemblies of molecules as advanced materials or smart materials [15]. Without question, the inspiration for this exciting development comes from the biological world, where, for example, the lipid bilayer of cell membranes was among the foremost self-assembling systems. In this connection it should be stated that many other researchers have also reported self-assembling systems such as Langmuir-Blodgett multilayers and liposomes [6-11, 16, 17]. The cogent reason that self-assembled BLMs are of scientific and practical interest is owing to the fact that most physiological activities involve some kind of lipid bilayer-based ligand-receptor contact interactions. Outstanding examples among these are ion sensing, antigen-antibody binding, hormone-synapse response, light conversion and detection, and gated channels, to name a few.

Many physiologically important processes are accompanied by charge transfer across the membrane and adjacent layers, and the related electrostatic phenomena are commonly recognized as a fundamental aspect of membrane biophysics. Both, charge transfer and binding to the surface depend on the electric field distribution at the membrane boundary, which in most cases is determined by the presence of charged lipid species. These circumstances have stimulated extensive use of planar lipid bilayers (BLM) and liposomes as model systems for studies of electrostatic phenomena at the membrane boundaries induced by inorganic ions and substances of biological interest [18, 19]. The distribution of electric field across the membrane interface in a microscopic scale is extremely complex and generally may be only approximated by smooth changes of potential in both directions perpendicular and parallel to membrane. The goal of electrochemical methods is to evaluate the difference between the potential in a reference point, usually taken as zero in the bulk of solution, and the potential averaged over a certain plane parallel to the membrane interface. One important imaginary plane (membrane surface) separates the membrane from the aqueous solution. The total potential drop across the interface, referred to as boundary potential, η_b, can be defined as the potential difference between the points in two phases, one of which is in the bulk of electrolyte, another is inside the membrane, near its hydrophobic core. This potential η_b is the sum of two parts (η_s and η_b). The first is the potential drop in the diffuse part of the electrical double layer and defined here as the surface potential, η_s. It is determined by the processes of surface ionization and screening by ions of the electrolyte. The Gouy Chapman-Stern (GCS) model provides an adequate description of these phenomena on the surface of biological membranes [19]. The other component of the boundary potential corresponds to the voltage drop across the interface and can be defined as the potential difference between two imaginary planes placed in different phases: in the aqueous solution immediately adjacent to the membrane surface and in the hydrophobic core inside the membrane. In contrast to the surface potential formed by the double layer, this interfacial potential drop is not accessible by any experimental approach because the energy of charge transfer between different phases includes not only the electrostatic but also an unmeasurable chemical component. This important point had been discussed in many monographs and textbooks [20-22]. Only the change of this component and therefore the variation of the total boundary potential can be monitored by external devices. The exact potential distribution in this region of the membrane is unknown and is presented in figure 1 as linear. As follows from many experiments [19, 23, 24] and numerous estimations [25], the interior of the membrane is more positive than the surrounding aqueous phase by about 200-300 mV. The physical nature of this potential difference reflects the molecular structure of the interface and may be ascribed to the orientation of dipole moments of lipid and water molecules or other moieties adsorbed or incorporated into lipid bilayer.

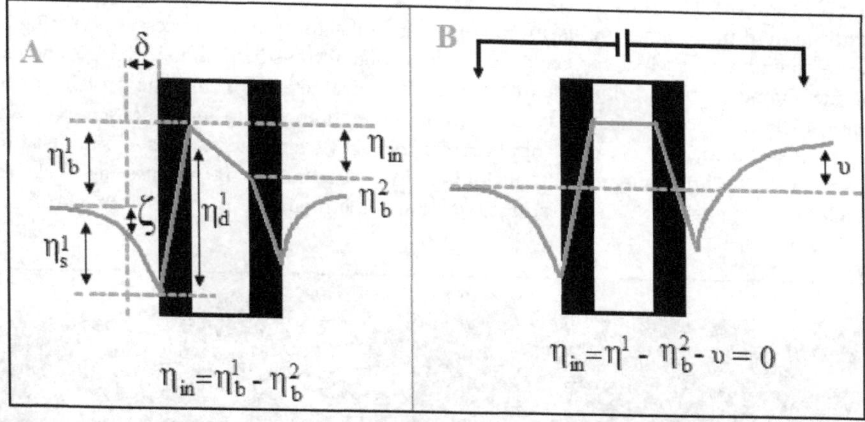

Figure 1. Electric potential distribution across the membrane under conditions of short-circuit (A) and intramembrane field compensation by an external voltage source (B). Hydrophobic core is indicated as light and polar regions as dark parts of the membrane. The vertical dashed line on the left shows the position of shear plane in the diffuse part of electrical double layer at distance 6 from the surface of membrane.

2. Basic concepts and definitions

The discovery of galvanic by L. Galvani in 1786 occurred simultaneously with his study of a bioelectrochemical phenomenon which was the response of excitable tissue to an electric impulse. E. du Bois-Reymond found in 1849 that such electrical phenomena occur at the surface of the tissue, but it was not until almost half a century later that W. Ostwald demonstrated that the site of these processes are electrochemical semipermeable membranes. In the next decade, research on semipermeable membranes progressed in two directions-in the search for models of biological membranes and in the study of actual biological membranes. The search for models of biological membranes led to the formation of a separate branch of electrochemistry, i.e. membrane electrochemistry. The most important results obtained in this field include the theory and application of ion-exchanger membranes and the discovery of ion-selective electrodes (including glass electrodes) and bilayer lipid membranes. The study of biological membranes led to the conclusion that the great majority of the processes in biological systems occur at cell and organelle membranes. The electrochemical aspects of this subject form the basis of bioelectrochemistry, dealing with the processes of charge separation and transport in biological membranes and their models, including electron and proton transfer in cell respiration and photosynthesis as well as ion transport in the channels of excitable cells. The electrokinetic phenomena (electrical double layer, interfacial tension of cells and organelles, cell membrane extension and contraction, etc.) also belong to this field.

3. Model systems of the biological membrane

Artificial lipid membranes are useful models to gain insight into the processes occurring at the cell membrane, such as molecular recognition, signal transduction, ion transport across the membrane. These membranes are often used to characterize membrane proteins or to study membrane active substances [26-28]. These membranes with incorporated receptors have a great potential in biosensor applications [29-33]. Various methods have been used to create artificial lipid membranes including free suspended membranes as well as membranes supported on a solid surface. Bellow is the model used to create this artificial lipid membranes.

Figure 2. Modern view of biological membranes (Picture generated by H. Seeger from Monte Carlo simulations and kindly provided by T. heimburg, NBI Copenhagen).

3.1. Liposomes

Liposomes, or lipid vesicles, are spherical structures in which an aqueous volume is enclosed by one or several lipid bilayers. They are usually made from phospholipids, which form energy-favorable structures in an aqueous solution due to the hydrophilic and hydrophobic interactions. Depending on the size and the number of bilayers, liposomes are classified as large multilamellar vesicles (MLV's) or large and small unilamellar vesicles (LUV's and SUV's). The size of unilamellar liposomes vary between 20 nm and 500 nm and the thickness of one lipid bilayer is about 4 nm. The structure of liposomes makes it possible to either encapsulate water-soluble molecules in the water interior of the liposome or immobilise molecules within the lipid membrane. Liposomes can be modified in a desired manner through the choice of membrane components and this has made them attractive as model system for cell membrane. In addition liposomes are frequently used as a delivery system for anticancer agents, increasing the effectiveness and circulation time of drugs. It is also possible to target specific cells by attaching an appropriate molecule at the liposome surface that binds specifically to the receptor site [34, 35]

3.2. Black lipid membranes (BLM)

In 1962 Müller et al pioneered the work of using black lipid membranes (BLM) as recognition elements. Unlike liposomes, BLMs are originally formed by spreading a lipid solution in a small hole (Ø = 0.5mm) of a wall separating two aqueous compartments. Evaporation or diffusion of the lipid solvent leads to thinning of the film to its final bilayer state. Black lipid membranes are very suitable for electro-chemical measurements, since there is easy electrochemical access to both sides of the membrane. They have the ability to control the constituents of each side of the membrane. In addition, there is no perturbing surface, and both the membrane and incorporated functionalities are likely to be close to their native state. The physical stability of these BLMs is very low and much effort has been put on improving the stability of this particular model membrane [36-38]. The residues left from solvent inside the bilayer have affected material properties of the bilayer, such as thickness [39], elasticity [40], and electrical properties [41]. The conduction of incorporated ion-channels is affected as well [42]. It is impossible to form large area BLMs because of the size limitation [43]. Nonetheless they have been extremely valuable in the history of membrane research.

3.3. Supported bilayer lipid membranes (sBLM)

An alternative to the BLM is to use lipid bilayers immobilized on a solid support. These membranes can be prepared on various surfaces, i.e., glass, silicon, mica, or on gold surfaces. The standard methods of preparing supported lipid membranes on planar solid surface are the Langmuir-Blodgett (LB) transfer and liposome spreading techniques. The major advantage of this membrane is its attachment to a solid support, resulting in a long-term and high mechanical stability. Solid supported membranes can be accessed by a variety of sensitive surface analysis tools such as surface Plasmon resonance spectroscopy, quartz crystal microbalance, scanning probe microscopy, as well as electrochemical measurements. But their close surface proximity between the artificial membrane and the bare solid surface onto which it is deposited restricts or even prevents the incorporation of large transmembrane-spanning proteins. The membrane-substrate distance is not sufficiently large to avoid direct contact between transmembrane proteins incorporated in the membrane and the solid surface. No ion transport can be detected with such membrane. Because embedded transmembrane proteins often have hydrophilic sections that protrude outside the lipid bilayer and may become immobile or denatured upon contact with the solid support. The lack of a well-defined ionic reservoir on the substrate side of the bilayer is a major drawback when studying membrane transport through ion carriers and channels. Hybrid bilayers have been developed to improve some of the above-mentioned shortcomings of the solid supported BLMs. In such sBLMs, the first monolayer is typically an alkanethiol monolayer, covalently attached to a metal substrate, onto which a phospholipids monolayeris deposited either by LB transfer or vesicle fusion. Such thiol/lipid solid supported BLMs are stable in air and constitute an improved barrier towards charge transfer [44,45]. However, the rigidity of alkanethiol/phospholipids solid supported BLMs is much higher than that of fluid biological membranes. In addition, the structure of thiol-based solid supported BLMs prevents the formation of a water layer between the bilayer and the metal support. Because of these conformational restraints, they are unsuitable for the incorporation of integral proteins and studies of ion transport through ionphores [46].

3.4. Polymer-supported bilayer lipid membranes (psBLM)

Attempts have been made to separate the membrane from the solid substrate by polymer cushions that rest on the substrate and support the membrane. Strategies include attachment to water-rich gels [47, 48], linking with covalently bound spacers [49-53] and whole spanning membranes [54]. A water layer is formed between the support and the hydrophilic headgroups. Tamm [55] and co-worker reported on the successful formation of tethered polymer-supported planar lipid bilayers. In their work a linear polyethyleneglycol (PEG) polymer was attached at its two ends to the substrate and a lipid molecule, respectively. Polymer-supported BLMs were developed in order to combine the most benefits of unsupported BLMs and solid supported BLM, such as bilayer fluidity and stability, accessibility to various characterization methods, and the possibility of incorporation and investigation of membrane proteins. Polymer-supported BLMs are stable and the presence of a thin, lubricating water layer between the substrate and the inner monolayer allow the incorporation and characterization of proteins. Lower impedance of such membranes limits the application of polymer-supported BLMs for biosensor purposes. The sealing property of such membrane is not comparable to the BLMs. The defects of the membrane may also prevent the incorporation of channel peptides and proteins.

3.5. Tethered bilayer lipid membranes (tBLM)

In an attempt to overcome the drawbacks of psBLM, sBLM and BLM, while preserving the stability provided by the covalently bound alkanethiol monolayer, so-called thiolipids have been synthesized [52, 56-59]. The thiolipids are composed of lipid derivatives, extended at their polar headgroups by hydrophilic spacers, which terminate in a thiol or disulfide group for covalent binding to the substrate. These molecules interact with gold surfaces, thus forming self-assembled monolayers hydrophobic to the outside. Once exposed to a suspension of liposomes, they tend to fusion and form lipid bilayers tethered to the gold surface by the hydrophilic spacer. Vogel and co-workers [56] were the first to introduce this concept. They synthesized so-called thiolipids with a head group composed of [60-62] ethyleneglycol units acting as a hydrogel and a thiol end group for covalent linkage to the gold surface. Bilayers were formed by depositing a second monolayer of different phosphocholines by the detergent dilution technique. Such bilayers exhibit a very high membrane resistance. Steinem et al [68] used different spacer and vesicle fusion technique to form a bilayer membrane with 0.8-1.0 μFcm^{-2} capacitance. Knoll and co-workers [58, 63] were the first to introduce to use oligopeptide sequences as hydrophilic spacers attached to the head group of lipid, resulting in the formation of peptide-tethered bilayers. Cornell and coworkers [29, 52, 64-65] were the first using half-membrane spanning tether lipids with benzyl disulphide (DPL) and synthetic archaea analogue full membrane spanning lipids with phytanolychains to stabilize the structure and polyethyleneglycol units as a hydrophilic spacer. Bilayer formation was achieved by immersion of a gold electrode in an ethanolic solution of the lipid mixture for the outer leaflet.

4. Membrane potentials

One of the property of electrochemical membranes is the formation of an electric potential difference ($\Delta\eta_{mem}$) between the two sides of the membrane and this is termed the membrane potential, and this quantity is determined by subtracting the left phase electric potential from right phase electric potential,

$$\Delta\eta_{mem} = \eta(2) - \eta(1) \tag{1}$$

For cell membranes, the intracellular liquid is usually denoted as solution 2, while solution 1 is the extracellular liquid.

The formation of a membrane potential is connected with the presence of an electrical double layer at the surface of the membrane. For a thick, compact membrane, an electrical double layer is formed at both interfaces. The electrical double layer at a porous membrane is formed primarily in the membrane pores. The electrical double layer at thin membranes is formed on both membrane surfaces. It is formed by fixed ions on the surface of the membrane and the diffuse layer in the electrolyte, by considering the simple case where both sides of the membrane are in contact with a solution of symmetrical electrolyte of the form MX in a single solvent and the membrane is permeable for only one ionic species (cation or anion only). In equilibrium its electrochemical potential in both solutions adjacent to the membrane has the same value. Thus if the membrane is permeable for cation M^{z+} the equation is

$$\Delta\eta_{mem} = \frac{RT}{zF} Ln\frac{a_{M(1)}}{a_{M(2)}} \tag{2}$$

and if the membrane is permeable for anion X^{z-} the equation is:

$$\Delta\eta_{mem} = \frac{RT}{zF} Ln\frac{a_{X(2)}}{a_{X(1)}} \dots \tag{3}$$

The membrane potential expressed by Eqs (2 and 3) is termed the **Nernst membrane potential** as it originates from the analogous ideas as the Nernst equation of the electrode potential and the equation of the Nernst potential at ITIES (Interface between Two Immiscible Electrolyte Solutions) of the form

$$E = \mathrm{E}^0 - \frac{RT}{nF} Ln\, Q \dots \tag{4}$$

Q is the reaction quotient.

Consider a system in which both solutions contain various ions for which the membrane is permeable and one type of ion that cannot pass through the membrane. The membrane is permeable for the solvent molecules to maintain the osmotic pressure across the two sides. The equilibrium conditions for the diffusible ions are

$$\boldsymbol{\mu'_i(1, P_1) = \mu'_i(2, P_2)} \tag{5}$$

This condition expresses the fact that the two solutions are under different pressures, p_1 and p_2, as a result of their different osmotic pressures. An analogous equation cannot be written for the non-diffusible ion as it cannot pass through the membrane and the 'equilibrium' concentrations cannot be established.

Consider dilute solutions, where it is possible to set $p_1 = p_2$ then the electrochemical potentials in Eq (5) are expanded in the usual manner, yielding for the diffusible cation

$$\boldsymbol{Ln \frac{a_{M(2)}}{a_{M(1)}} = -\frac{(Z_+)F}{RT} \Delta\eta_{mem}} \cdots \tag{6}$$

and for the diffusible anion,

$$\boldsymbol{Ln \frac{a_{X(2)}}{a_{X(1)}} = -\frac{(Z_-)F}{RT} \Delta\eta_{mem}} \tag{7}$$

Rearranging these equations yielding,

$$\boldsymbol{\left[\frac{a_{M(2)}}{a_{M(1)}}\right]^{1/z_+} = exp\left[-\frac{(Z_+)F}{RT} \Delta\eta_{mem}\right]^{1/z_+} = \lambda} \tag{8}$$

i.e. for univalent, divalent, etc., cations and anions

$$\boldsymbol{\frac{a_{M(2)}}{a_{M(1)}} = \left[\frac{a_{M(2)}}{a_{M(1)}}\right]^{1/2} = \left[\frac{a_{M(2)}}{a_{M(1)}}\right]^{1/3} = \lambda} \tag{9}$$

$$\boldsymbol{\frac{a_{X(2)}}{a_{X(1)}} = \left[\frac{a_{X(2)}}{a_{X(1)}}\right]^{1/2} = \left[\frac{a_{X(2)}}{a_{X(1)}}\right]^{1/3} = \lambda} \tag{10}$$

The constant $\boldsymbol{\lambda}$ is termed the **Donnan distribution coefficient**.

In the simple case of a diffusible, univalent cation M^+ and anion X^- and non-diffusible anion N^- present in phase 2, the condition of electroneutrality in phase 2 gives,

$$[\boldsymbol{M}^+]_{(2)} = [\boldsymbol{X}^-]_{(2)} + [N^-]_{(2)} \cdots \tag{11}$$

and in phase 1 gives,

$$[\boldsymbol{M}^+]_{(1)} = [\boldsymbol{X}^-]_{(1)} \cdots \tag{12}$$

For dilute solutions, activities can be replaced by concentrations by taking activity coefficient equal to one, and then equations (9 and 10) give,

$$[M^{+}]_{(1)}\,[X^{-}]_{(1)} = [M^{+}]_{(2)}\,[X^{-}]_{(2)} \cdots \quad .(13)$$

The Donnan distribution constant $\boldsymbol{\lambda}$ from the combination of equations (11-13) is,

$$\boldsymbol{\lambda} = \left(\frac{[X^{-}]_{(2)}[N^{-}]_{(2)}}{[X^{-}]_{(1)}}\right) \quad (14)$$

A more exact solution of the equilibrium conditions (Eq. 5) must consider that the standard term $\boldsymbol{\mu}°$ depends on the pressure, which is different in the two solutions,

$$\boldsymbol{\mu^{0}(T,P) = \mu^{*}(T,P) = \int_{0}^{P} v dP} \ldots \quad (15)$$

where $\boldsymbol{\mu}^*$ is the limiting value of $\boldsymbol{\mu}°$ at P →0 and v is the molar volume of the component at pressure P. If the volume v is considered to be independent of the pressure (more accurate calculations employ a linear dependence) then,

$$\boldsymbol{\mu^{0}(T,P) = \mu^{*}(T) + \nu P.} \quad (16)$$

The Donnan potential at dilute solution $\Delta\eta_D = \Delta\eta_{mem}$ can be found as,

$$\boldsymbol{\Delta\eta_{D} = -\frac{RT}{F} Ln\, \lambda} \ldots \quad (17)$$

The Donnan potentials contain the individual ionic activities and cannot be measured by using a purely thermodynamic procedure. In the concentration range where the Debye-Hûckel limiting law is valid, the ionic activities can be replaced by the mean activities. The membrane potentials are measured by constructing a cell with a semipermeable membrane separating solutions 1 and 2:

$$\boldsymbol{Ag|\, AgCl, sat. KCl|\, Solution\ 1|\, Solution\ 2|\, Sat. KCl, AgCl|\, Ag}.. \quad (18)$$

5. Evidence for electronic processes in Membranes

The origin of the concept of electronic processes in membranes and related systems was first reviewed in 1971, in which the phenomenon known as "electrostenolysis" was stressed. In the language of membrane electrochemistry, electrostenolysis simply means that a reduction reaction takes place on one side of the membrane where the positive electrode is situated and the oxidation occurs on the other side of the membrane. Although electronic processes in BLM in the dark were mentioned in 1970, no conclusive evidence has been shown. This is because of the fact that an unmodified B LM is an excellent insulator (resistivity >10^{15} ohms) incapable of either ionic or electronic conduction.

In order to demonstrate electronic conduction in BLM it seems that the membrane must, first of all, be modified to function as a redox or semiconductor electrode. Secondly, an appropriate method must be found for studies of electronic processes that can be applied to the BLM system. In the following paragraphs evidence of electronic processes in membranes, in particular in BLMs in the dark is described.

6. Electron-conducting BLMs

Consider an electron-conducting bilayer lipid membrane (BLM) is separating two aqueous solutions containing different redox couples, let the left side referred as the outside "out" (or Aqueous Solution 1) and the right side referred as the inside "in" (or Aqueous Solution 2). To facilitate the discussion, let's hold the outside constant and consider the interfacial electron transfer reaction on the inside,

$$\boldsymbol{D} \underset{k_f}{\overset{k_b}{\Leftarrow\Rightarrow}} \boldsymbol{D}^{+} - \boldsymbol{n}\boldsymbol{e}^{-} \tag{19}$$

where D and D^+ denote an electron donor in its reduced and oxidized form, respectively, whose concentrations are $[D^+]_2$ and $[D]_2$ in the bulk phase. The corresponding interfacial concentrations are $[D^+]_2^s$ and $[D]_2^s$. The forward and backward heterogeneous rate constants are k_f and k_b, respectively. The observed current is then given by

$$\boldsymbol{I}_{net} = \boldsymbol{nFA}\left(\boldsymbol{k}_f[\boldsymbol{D}]_{2-}\boldsymbol{k}_b[\boldsymbol{D}^{+}]_2\right) \tag{20}$$

n is the number of moles of the electrons changed through the process, F is the faraday constant and A is the area of the electrode.

Let the observed membrane potential denoted by ($\Delta\eta_{mem}$), to explain this it is best to assume the membrane is behave like an ideal electron conductor just like platinum wire. From thermodynamics, the overall free energy change associated with an electrical cell reaction is given by

$$\boldsymbol{\Delta G} = -\boldsymbol{nFDE}\ \ldots \tag{21}$$

where E is the electromotive force of the cell. Consider the BLM system as an electrical cell, each solution/membrane interface is assumed to behave as a redox electrode. From eq. (21) the individual free energy terms are denoted by Ğ. Thus, as indicated on the left side of the BLM,

$$\mathbf{A} + \mathbf{e} \longrightarrow \mathbf{A}^{-}$$

Then

$$\breve{G}_A + \breve{G}_e = \breve{G}_{A^-} \tag{22}$$

Similarly, on the right side,

$$\mathbf{D} - \mathbf{e} \longrightarrow \mathbf{D}^{+}$$

Then

$$\breve{G}_D - \breve{G}_e = \breve{G}_{D^+} \tag{23}$$

Here electrons involved are considered as one of the reactants. Since

$$\breve{G}_i = \breve{G}^0 + \boldsymbol{RT}\,\boldsymbol{Ln}\,\boldsymbol{a}_i\ \ldots \tag{24}$$

Taking the activity coeffiecent is equal to one, then

$$\breve{G}_i = \breve{G}^0 + \boldsymbol{RT\,Ln}\,[i] \tag{25}$$

Where $\breve{G}^0$ is a constant, R and T have the usual significance, [i] is the concentration of the redox species when the activity coefficient is taken to the unity. The free energy of the electron $\breve{G}_e$ for the outer solution/membrane interface, is given by

$$\breve{G}_e = -\boldsymbol{FE_{out}} = \breve{G}_{A^-} - \breve{G}_A = (\breve{G}^0_{A^-} - \breve{G}^0_A) + \boldsymbol{RT\,Ln}\,\frac{[A^-]}{[A]} \ldots \tag{26}$$

where E_{out} is the electrode potential.

and the free energy of the electron $\breve{G}_e$ for the inner solution/membrane interface, we have

$$\breve{G}_e = -\boldsymbol{FE_{in}} = \breve{G}_D - \breve{G}_{D^+} = (\breve{G}^0_D - \breve{G}^0_{D^+}) + \boldsymbol{RT\,Ln}\,\frac{[D]}{[D^+]} \tag{27}$$

At equilibrium, the free energy difference of the electron across the BLM is equal to

$$\Delta\eta_{mem} = \left(\boldsymbol{E^0_{(A/A^-)}} - \boldsymbol{E^0_{(D^+/D)}}\right) + \frac{\boldsymbol{RT}}{\boldsymbol{F}}\boldsymbol{Ln}\frac{[A][D]}{[A^-][D^+]} \tag{28}$$

In case of $[A^-] = [A]$ and $[D^+] = [D]$ the logarithmic part will be equal to one, then $\Delta\eta_{mem}$ will be

$$\Delta\eta_{mem} = \left(\boldsymbol{E^0_{(A/A^-)}} - \boldsymbol{E^0_{(D^+/D)}}\right) = \boldsymbol{E_{redox^0}} \ldots \tag{29}$$

7. Examples of biological membrane processes, processes in the cells of excitable tissues

The transport of information from sensors to the central nervous system and of instructions from the central nervous system to the various organs occurs through electric impulses transported by nerve cells. These cells consist of a body with star-like projections and a long fibrous tail called an axon.

While in some mollusks the whole membrane is in contact with the intercellular liquid, in other animals it is covered with a multiple myeline layer which is interrupted in definite segments (nodes of Ranvier). The Na^+,K^+-ATPase located in the membrane maintains marked ionic concentration differences in the nerve cell and in the intercellular liquid. The relationship between the electrical excitation of the axon and the membrane potential was clarified by A. L. Hodgkin and A. P. Huxley.

If the axon is not excited, the membrane potential

$$\Delta\eta_{mem} = \eta_{in} - \eta_{out} \tag{30}$$

has a rest value of about -90 mV. When the cell is excited by small square wave current impulses, a change occurs in the membrane potential roughly proportional to the magnitude of the excitation current impulses. If current flows from the interior of the cell to the exterior, then the absolute value of the membrane potential increases and the membrane is hyperpolarized. Current flowing in the opposite direction has a depolarization effect and the absolute potential value decreases.

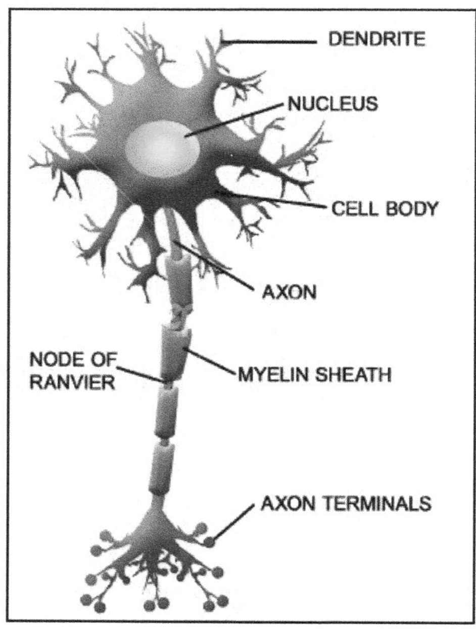

Figure 3. Structure of neuron cell.

When the depolarization impulse exceeds a certain 'threshold' value, the potential suddenly increases. The characteristic potential maximum is called a spike, figure 4, and its height no longer depends on a further increase in the excitation impulse. Sufficiently large excitation of the membrane results in a large increase in the membrane permeability for sodium ions so that, finally, the membrane potential almost acquires the value of the Nernst potential for sodium ions ($\Delta\eta_{mem}$ = +50 mV).

A potential drop to the rest value is accompanied by a temporary influx of sodium ions from the intercellular liquid into the axon. If the nerve is excited by a subthreshold current impulse, then a change in the membrane potential is produced that disappears at a small distance from the excitation site (at most 2 mm). A spike produced by a threshold or larger current impulse produces further excitation along the membrane, yielding further spikes that are propagated along the axon. As already pointed out, sodium ions are transferred from the intercellular liquid into the axon during the spike. This gradual formation and disappearance of positive charges corresponds to the flow of positive electric current along the axon. An adequate conductance of thick bare cephalopod axons allows the flow of sufficiently strong currents. In myelinized axons of vertebrates a much larger charge is formed (due to the much higher density of sodium channels in the nodes of Ranvier) which moves at high speed through much thinner axons than those of cephalopods. The myelin sheath then insulates the nerve fibre, impeding in this way the induction of an opposite current in the intercellular liquid which would hinder current flow inside the axon.

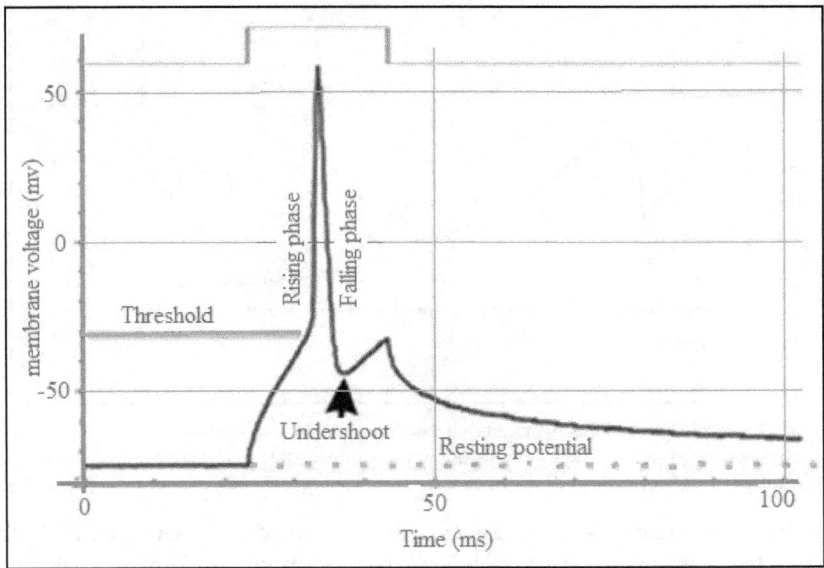

Figure 4. The characteristic potential maximum.

Another way to study membrane electrochemistry is the so-called voltageclamp method which is based on polarizing the membrane by using a fourelectrode potentiostatic arrangement. In this way, Cole [66] showed that individual currents linked to selective ion transfer across the membrane are responsible for impulse generation and propagation.

A typical current-time curve is shown in figure (5), obviously, the membrane ion transfer is activated at the start, but after some time it becomes gradually inhibited. The ion transfer rate typically depends on both the outer (bathing) and inner solution (the inside of a cephalopod membrane as much as 1 mm thick can be rinsed with an electrolyte solution without affecting its activity).

The assumption that the membrane currents are due to ion transfer through ion-specific channels was shown correct by means of experiments where the channels responsible for transfer of a certain ionic species were blocked by specific agents. Thus, the sodium transfer is inhibited by the toxin, tetrodotoxin, while the transport of potassium ions is blocked by the tetraalkylammonium ion with three ethyl groups and one longer alkyl group, such as a nonyl. The effect of toxins on the ion transport across the axon membrane, which occurs at very low concentrations, has led to the conclusion that the membrane contains ion-selective channels responsible for ion transport. This assumption was confirmed by analysis of the noise level in ionic currents resulting from channel opening and closing.

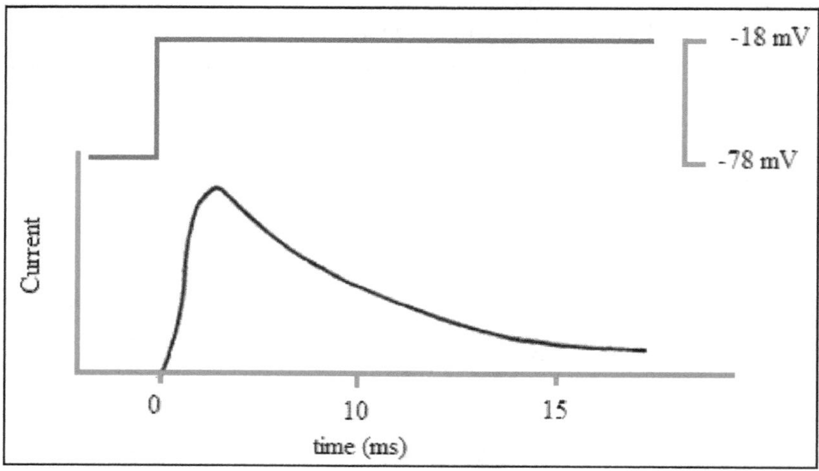

Figure 5. Time dependence of the membrane current. Since the potassium channel is blocked the current corresponds to sodium transport. The upper the represents the time course of the imposed potential difference. (According to W. ulbricht)

In this regards we should mentioned here that the structure of the Na and K channels differ from type to type of animal classes, recently from our study on the conformational changes of Na^+,K^+-ATPase from some type of animal classes using stopped-flow techniques have revealed major differences in the kinetic mechanisms (and hence enzyme structure) of mammalians and non-mammalians enzymes [69,70], in the absence of ATP appear that the mammalian Na^+,K^+-ATPase exists in a diprotomeric state $(\alpha\beta)_2$, with protein-protein interactions between the α-subunits causing an inhibition of the transition, while binding of ATP to any of the enzyme conformational states induced the dissociation of the diprotomer into separate $\alpha\beta$ protomers and relief the pre-existing inhibition, non-mammalians exhibits no effect of ATP binding on the enzyme at all concentrations indicating a mono-protomeric structure ($\alpha\beta$ protomers).

Because of the very low extraneous noise level, the patch-clamp method permits the measurement of picoampere currents in the millisecond range figure (6).

Ion channels of excitable cells consist of a narrow pore, of a gate that opens and closes the access to the pore, and of a sensor that reacts to the stimuli from outside and issues instructions to the gate. The outer stimuli are either a potential change or binding of a specific compound to the sensor. The nerve axon sodium channel was studied in detail (in fact, as shown by the power spectrum analysis, there are two sorts of this channel: one with fast opening and slow inactivation and the other with opposite properties).

The rate of transport of sodium ions through the channel is considerable, when polarizing the membrane with a potential difference 4-60mV a current of approximately 1.5 pA flows through the channel which corresponds to 6 x 10^6 Na^+ ions per second, practically the same value as with the gramicidin A channel.

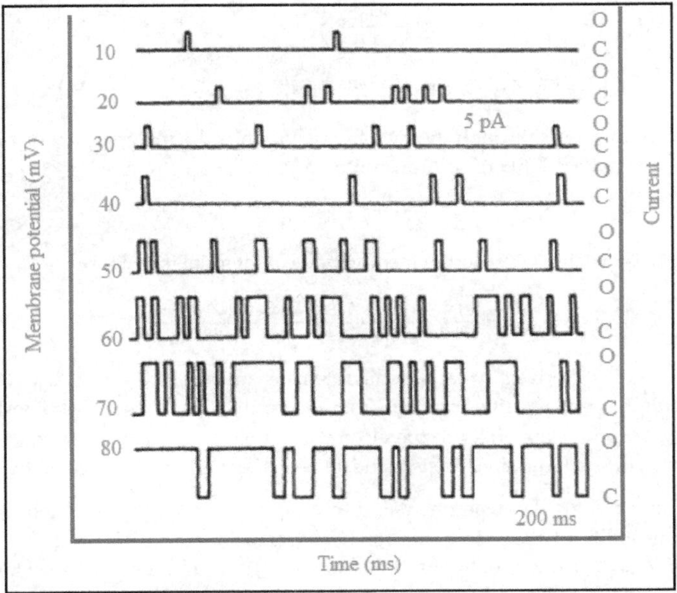

Figure 6. Joint application of patch-clamp and voltage-clamp methods to the study of single potassium channel present in the membrane of spinal-cord neuron cultivated in the tissue culture. The values indicated before each curve is potential differences imposed on the membrane. The ion channel is either closed (C) or opens (O). (According to B. Hille)

The sodium channel is only selective but not specific for sodium transport. It shows approximately the same permeability to lithium ions, whereas it is roughly ten times lower than for potassium. The density of sodium channels varies among different animals, being only 30 μm^{-2} in the case of some marine animals and 330 μm^{-2} in the squid axon, reaching 1.2x$10^4 \mu m^{-2}$ in the mammalian nodes of Ranvier. The potassium channel mentioned above, figure (6), is more specific for K^+ than the sodium channel for Na^+ being almost impermeable to Na^+.

In deriving a relationship for the resting potential of the axon membrane it will be assumed that, in the vicinity of the resting potential, the frequency of opening of a definite kind of ion channel is not markedly dependent on the membrane potential. It will be assumed that the resting potential is determined by the transport of potassium, sodium and chloride ions alone. The constants k_i^θ are functions of the frequency of opening and closing of the gates of the ion-selective channels. The solution to this problem will be based on analogous assumptions to those employed for the mixed potential. The material fluxes of the individual ions are given by the equations

$$J_{K^+} = k^{\theta}_{K^+} C_{K^+}(1) exp\left(\frac{-F\Delta\eta_{mem}}{2RT}\right) - k^{\theta}_{K^+} C_{K^+}(2) exp\left(\frac{F\Delta\eta_{mem}}{2RT}\right) \quad (31)$$

$$J_{Na^+} = k^{\theta}_{Na^+} C_{Na^+}(1) exp\left(\frac{-F\Delta\eta_{mem}}{2RT}\right) - k^{\theta}_{Na^+} C_{Na^+}(2) exp\left(\frac{F\Delta\eta_{mem}}{2RT}\right).. \quad (32)$$

$$J_{Cl^-} = k^{\theta}_{Cl^-} C_{Cl^-}(1) exp\left(\frac{F\Delta\eta_{mem}}{2RT}\right) - k^{\theta}_{Cl^-} C_{Cl^-}(2) exp\left(\frac{-F\Delta\eta_{mem}}{2RT}\right). \quad (33)$$

At rest, no current passes through the membrane and thus the material flux of chloride ions compensates the material flux of sodium and potassium ions, so that,

$$J_{Na^+} + J_{K^+} = J_{Cl^-} \quad (34)$$

Equations (31, 32, 33 and 34) yield the membrane rest potential in the form

$$\Delta\eta_{mem} = \frac{RT}{F} Ln \frac{k^{\theta}_{K^+} C_{K^+}(1) + k^{\theta}_{Na^+} C_{Na^+}(1) + k^{\theta}_{Cl^-} C_{Cl^-}(2)}{k^{\theta}_{K^+} C_{K^+}(2) + k^{\theta}_{Na^+} C_{Na^+}(2) + k^{\theta}_{Cl^-} C_{Cl^-}(1)} \cdots \quad (35)$$

Ion transport is characterized by conditional rate constants $k^{\Theta}_{K^+}$, $k^{\Theta}_{Na^+}$ and $k^{\Theta}_{Cl^-}$ which can be identified with the permeabilities of the membrane for these ions. These relationships can be improved by including the effect of the electrical double layer on the ion concentration at the membrane surface. Equation (35) is identical with the relationship derived by Katz [67].

It satisfactorily explains the experimental values of the membrane rest potential assuming that the permeability of the membrane for K^+ is greater than for Na^+ and Cl^-, so that the deviation of $\Delta\eta_{mem}$ from the Nernst potential for K^+ is not very large. However, the permeabilities for the other ions are not negligible. In this way the axon at rest would lose potassium ions and gain a corresponding concentration of sodium ions. This does not occur because of the action of Na^+,K^+-ATPase, transferring potassium ions from the intercellular liquid into the axon and sodium ions in the opposite direction, through hydrolysis of ATP. When the nerve cells are excited by an electric impulse (either natural from another nerve cell or another site on the axon, or artificial from an electrode), the membrane potential changes, causing an increase in the frequency of opening of the gates of the sodium channels. Thus, the flux of sodium ions increases and the membrane potential is shifted towards the Nernst potential value determined by sodium ions, which considerably differs from that determined by potassium ions, as the concentration of sodium ions in the extracellular space is much greater than in the intracellular space, while the concentration ratio of potassium ions is the opposite. The potential shift in this direction leads to a further opening of the sodium gates and thus to 'autocatalysis' of the sodium flux, resulting in a spike which is stopped only by inactivation of the gates. In spite of the fact that the overall currents flowing across the cell membranes consist of tiny stochastic fluctuating components, the resulting dependences, as shown in figure (5), are smooth curves and can be used for further analysis (the situation is, in fact, analogous to most of phenomena occurring in nature). Thus, the formation of a spike can be shown to be a result of gradual opening and closing of many potassium and sodium channels figure (7).

According to figure (3) the nerve cell is linked to other excitable, both nerve and muscle, cells by structures called, in the case of other nerve cells, as partners, synapses, and in the case of striated muscle cells, motor endplates (***neuromuscular junctions***). The impulse, which is originally electric, is transformed into a chemical stimulus and again into an electrical impulse. The opening and closing of ion-selective channels present in these junctions depend on either electric or chemical actions. The substances that are active in the latter case are called ***neurotransmitters***. A very important member of this family is acetylcholine which is transferred to the cell that receives the signal across the postsynaptic membrane or motor endplate through a specific channel, the nicotinic acetylcholine receptor. This channel has been investigated in detail, because, among other reasons, it can be isolated in considerable quantities as the membranes of the cells forming the electric organ of electric fish are filled with this species.

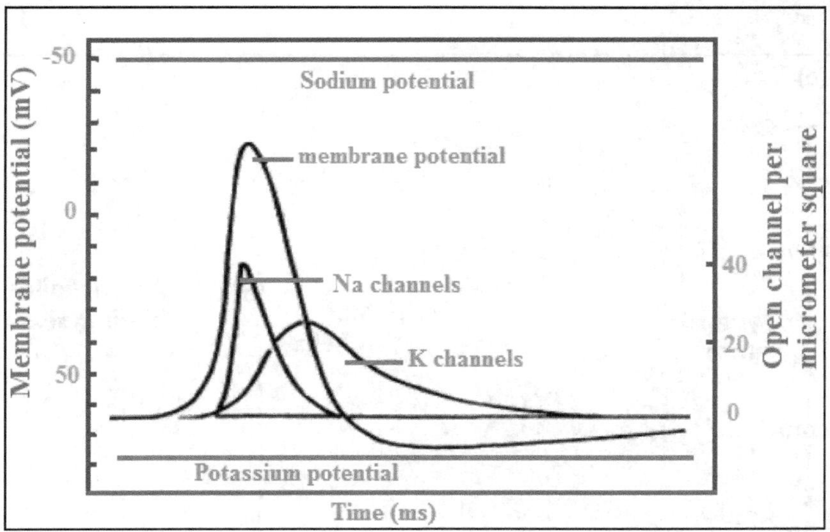

Figure 7. A hypothetical scheme of the time behavior of the spike linked to the opening and closing of sodium and potassium channels. After longer time intervals a temporary hyperpolarization of the membrane is induced by reversed transport of potassium ions inside the nerve cell. (According to A.L. Hodgkin and A.F. huxley)

Acetylcholine, which is set free from vesicles present in the neighborhood of the presynaptic membrane, is transferred into the recipient cell through this channel figure (8). Once transferred it stimulates generation of a spike at the membrane of the recipient cell. The action of acetylcholine is inhibited by the enzyme, acetylcholinesterase, which splits acetylcholine to choline and acetic acid. Locomotion in higher organisms and other mechanical actions are made possible by the striated skeletal muscles. The basic structural unit of muscles is the muscle cell-muscle fibre which is enclosed by a sarcoplasmatic membrane. This membrane invaginates into the interior of the fibre through transversal tubules which are filled with the intercellular liquid. The inside of the fibre consists of the actual sarcoplasm with inserted mitochondria, sarcoplasmatic reticulum and minute fibres called myofibrils, which are the organs of muscle contraction and relaxation. The membrane of the sarcoplasmatic reticulum contains Ca^{2+}-ATPase, maintaining a concentration of calcium ions in the sarcoplasm of the relaxed muscle below 10^{-7}mol.dm^{-3}. Under these conditions, the proteins, actin and myosin, forming the myofibrils lie in relative positions such that the muscle is relaxed. When a spike is transferred from the nerve fibre to the sarcoplasmatic membrane, another spike is also formed there which continues through the transverse tubules to the membrane of the sarcoplasmatic reticulum, increasing the permeability of the membrane for calcium ions by five orders within 1 millisecond, so that the concentration of Ca^{2+} ions in the sarcoplasm increases above 10^{-3} mol.dm^{-3}. This produces a relative shift of actin and myosin molecules and contraction of the muscle fibre, and after disappearance of the spike, Ca^{2+}-ATPase renews the original situation and the muscle is relaxed.

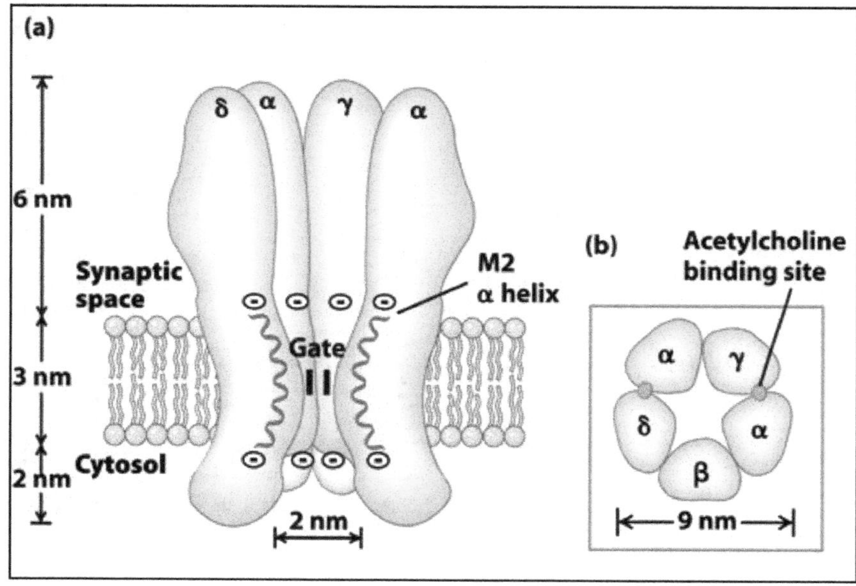

Figure 8. The nicotinic acetylcholine receptor in a membrane. The deciphering of the structure is based on X-ray diffraction and electron microscopy. (Molecular cell biology, sixth edition, 2008, W. H. freeman and company).

Abbreviations

BLMs bilayer lipid membranes
GCS Gouy Chapman-Stern model
MLV's multilamellar vesicles

LUV's	large unilamellar vesicles
SUV's	small unilamellar vesicles
BLM	Black lipid membranes
LB	Langmuir-Blodgett transfer
psBLM	Polymer-supported bilayer lipid membranes
tBLM	Tethered bilayer lipid membranes
ITIES	Interface between Two Immiscible Electrolyte Solutions
$\Delta\eta_{mem}$	membrane potential
T	Absolute temperature
R	Universal gas constant
F	Faraday constant
Z	Ion charge
a	Activity
η_b	boundary potential
η_s	surface potential
E	Electrode potential
E_0	Electrode standard potential
P	Pressure
λ	Donnan distribution coefficient
Q	Reaction quotient
N	non-diffusible anion
v	is the molar volume of the component at pressure P
$\Delta\eta_D$	Donnan potential at dilute solution
I_{net}	The observed current

Author details

Mohammed Awad Ali Khalid
Department of Chemistry, College of applied medical and Science,
University of Taif, Saudi Arabia
Department of Chemistry, Faculty of Science, University of Khartoum, Sudan

8. References

[1] K.L. Mittal and D.O. Shah (Eds.), Surfactants in Solution, Plenum Press. NY. 8 (1989) 133: II (1991) 61.
[2] M. Blank (Ed.) Biomembrane Electrochemistry, Adv. Chem, Series No. 235, American Chemical Society, Washington. DC. 1994.
[3] J.O'M. Bockris and S.U.M. Khan. Surface Electrochemistry, Plenum Press, NY. 1993.
[4] J. Wang, Analytical Electrochemistry, VCH Publishers, NY.1994
[5] H.T. Tien, J. Phys. Chem. 88 (1994) 3172: J. Electroanal. Chem. 174 (1984) 299
[6] A. ottova-Leitmannuva and H.T. Tien, Prog. Surf. Sci., 41 (1992) 337.

[7] S.G. Davison (Ed.) Progress in Surface Science, Pergamon Press, NY, 4 (1973): 19 (1985) 169: 23 (1986) 317:30 (1989) 1.
[8] A.D. Bangham. BioEssays, 17 (1995) 1081.
[9] P.-A. Ohlsson, T. Tjamhage, E. Iterbai, S. Lofas and G. Puu. Bioelectrochem. Bioenerg., 38 (1995) 137
[10] T. Vo-Dinh, Sensors Actuators B, 29 (1995) 183.
[11] F.M. Menger and K.D. Gabriels, Angew. Chem., 34 (1995) 2091.
[12] B.W. Koenig, S. Krueger, W.J. Orts, C.F. Majkrzak, N.F. Berk, J.V. Silverton and K. Gawrisch, Langmuir, 12 (1996) 1343.
[13] N. Marakami, S.S. Singh, V.P.S. Chauhan and M. Elziaga, Biochem., 34 (1995) 6046.
[14] Y.Q. Liang, Z.Q. Zhang, L.X. Wu, Y.C. Tian and H.D. Chen, J. Coll. Int. Sci., 178 (1996) 714.
[15] H.T. Tien, Mat. Sci. Eng. C. 3 (1995) 7; in Proc. 5th Int. Symp. On Bioelectronics and Molecular Electronic Devices, 28-30 November 1995, Okinawa, Japan. p. 81.
[16] P. Tancrede, P. Paquin, A. Houle and R.M. LeBlanc, J. Biochem. Biophys. Meth., 7 (1983) 299.
[17] T.D. Osborn, and P Yager, Langmair, 11 (1995) 8.
[18] G. Cevc and D.Marsh. Phospholipid Bilayers, Physical Principles and Models, Vol.5 (1987), New York, Willey-Interscience Publication, Cell Biology: A Series of Monographs. Bittar, E. E. (ed.)
[19] S. McLaughlin, Electrostatic Potentials at Membrane-Solution Interfaces, Vol.9, 71-144 (1977) New York. Current Topics Membranes and Transport. Bronnen, F. and Kleinzeller, A. (eds.)
[20] E.A. Guggenheim, Thermodynamics, Wiley, New York, (1950).
[21] A.W. Adamson, Physical Chemistry of Surfaces, Wiley&Sons, (1967).
[22] J.T. Davies and E.K. Rideal, Interfacial Phenomena, Academic Press, 2nd ed., New York, (1963).
[23] A.D. Pickar and R. Benz, J. Membrane Biol., 44 (1978) 353.
[24] K. Gawrisch, D. Ruston, J. Zimmerberg, A. Parsegian, R.P. Rand, and N. Fuller. Biophys. J., 61 (1992)1213.
[25] R.F. Flewelling and W.L. Hubbell., Biophys. J., 49 (1986) 541.
[26] A. Charbit, C. Andersen, J. Wang, B. Schiffler, V. Michel, R. Benz and M. Hofnung, Mol. Microbiol., 35 (2000), 777.
[27] T. K. Rostvtseva, T. T. Liu, M. Colombini, V. A. Parsegian and S. M. Bezrukov, Proc. Nat. Acad. Sci. USA, 97 (2000), 7819.
[28] M. Akeson, D. Branton, J. J. Kasianowics, E. Brandin and D. W Deamer, J. Biophys., 77 (1999), 3227.
[29] B. A. Cornell, V. L. Braach-Maksvytis, L. G. King, P. D. Osman, B. Raguse, L. Wieczorek and R. J. Pace, Nature, 387 (1997), 580.
[30] S. Heyse, T. Stora, E. Schmid, J. H. Lakey and H. Vogel, Biochim. Biophys. Acta, 1376 (1998), 319.
[31] K. Seifert, K. Fendler and E. Bamberg, Biophys. J., 64 (1993), 384
[32] T. Stora, J. H. Lakey and H. Vogel, Angew. Chem. Int. Ed. Engl., 38 (1999), 389

[33] M. Winterhalter, Colloids Surf. A, 149 (1999), 547.
[34] R. J. Banerjee, Biomater. Appl. 16 (2001), 3-21.
[35] D.D. Lasic, Trends Biotechnol., 1998, 16, 307-321
[36] H.T. Tien, and A. Ottova-Leitmannova, Elsevier, Amsterdam, (2003)
[37] J.D. Castillo, A. Rodriguez, C.A. Romero, and V. Sanchez, Science, 153 (1966) 185-188
[38] R.G. Ashcroft, H.G.L. Coster, D.R. Laver, and J.R. Smith, Biochmica et Biophysica Acta, 730 (1983) 231-238.
[39] R. Fettiplace, D.M. Andrews, and D.A. Haydon, J. Membrane Biol., 5 (1971) 277-
[40] T. Hianik, J.Miklovicova, A. Bajci, D. Chorvat, and V. Sajter, Gen. Physiol. Biophys., 3 (1984) 79-84.
[41] H.G.L. Coster, and D.R. Laver, Biochim. Biophys. Acta, 857 (1986) 95-104.
[42] B.M. Hendry, B.W. Urban, and D.A. Haydon, Biochim. Biophys. Acta, 513 (1978)106-116.
[43] O. Purrucker, H. Hillebrandt, K. Adlkofer, and M.Tanaka, Eletrochim. Acta, 47 (2001) 791-798
[44] A.L. Plant, Langmuir, 9 (1993) 2764-2767.
[45] S. Lingler, I. Rubinsten, W. Knoll, A. Offenhäusser, Langmuir, 13 (1997) 8085-7091.
[46] R. Guidelli, G. Aloisi, L. Becucci, A. Dolfi, M.R. Moncelli; F T. Buoninsegni, J. Electroanal. Chem., 504 (2001), 1-28.
[47] H.T. Tien, and A.L. Ottova, Colloids and Surfaces A. Physicochemical and Engineering Aspects, 149 (1999) 217-233.
[48] T. Ide, and T. Yanagida, Biochemical and Biophysical Research Communications, 265 (1999) 595-599.
[49] Y. Cheng, N. Boden, R.J. Bushby, S. Clarkson, S.D. Evans, P.F. Knowles, A. Marsh, and R.E. Miles, Langmuir, 14 (1998) 839-844.
[50] N. Bunjes, E.K. Schmidt, A. Jonczyk, F. Rippmann, D. Beyer, H. Ringsdorf, P. Gräber, W. Knoll, R. Naumann, Langmuir, 13 (1997) 6188-6194.
[51] B.A. Cornell, V.L.B. Braach-Maksvytis, L.G. King, P.D.J. Osman, B. Raguse, L. Wieczorek, R.J. Pace, Science, 387 (1997) 580-583.
[52] B. Raguse, V. Braach-Maksvytis, B.A. Cornell, L.G. King, P.D.J. Osman, R.J. Pace, L. Wieczorek, Langmuir, 14 (1998) 648-659.
[53] A.T.A. Jenkins, R.J. Bushby, N. Boden, S.D. Evans, P.F. Knowles, Q. Liu, R.E. Miles, S.D. Ogier, Langmuir, 14 (1998) 4675-4678
[54] C. Schmidt, M. Mayer, and H.Vogel, Angew. Chem. Int. Ed.,39 (2000)3137-3140
[55] V. Kiessling, L.K. Tamm, Biophys J., 47 (2003) 105-113.
[56] H. Lang, C. Duschl, H. Vogel, Langmuir, 10 (1994) 197.
[57] C. Steinem, A. Janshoff, J. Goossens, H-J. Galla, Bioelectrochem. Bioenergetics, 45 (1998) 17-26.
[58] R. Naumann, A. Jonczyk, R. Kopp, v. Esch, J. Ringsdorf, H. Knoll, W. Gräber, P. Angew., Chem., 34 (1995) 2056.
[59] L.M.Williams, S.D. Evans, T.M. Flynn, A. Marsh, P.F. Knowles, R. J. Bushby, N. Boden, Langmuir 13 (1997) 751.
[60] S.J. Singer, G.L. Nicolson, Science, 175 (1972) 720-731

[61] O.G. Mouristsen, In Life-as a matter of fate: The Emerging science of lipidomics; D. Dragomann, M. Dragomann, A.C. Elitzur, M.P. Silverman, J. Tuszynski, and H.D.Zeh, Eds., The Frontiers Collection, Springer, Germany, Vol.1 (2005) pp 27.

[62] A. Charbit, C. Andersen, J. Wang, B. Schiffler, V. Michel, R. Benz and M. Hofnung, Mol. Microbiol., 35 (2000) 777

[63] R. Naumann, E.K. Schmidt, A. Jonczyk, K. Fendler, B. Kadenbach, T. Lieberman, A. Offenhäussser, W. Knoll, Biosens Bioelectronics, 14 (1999) 651-662

[64] G.E. Woodhouse, L.G. King, L. Wiezcorek, B.A. Cornell, Faraday Disc., 111 (1998) 247.

[65] B.A. Cornell, G. Krishna, P.D. Osman, R.J. Pace, L. Wieczorek, Biochemical Soc. Transactions, 29 (2001) 613.

[66] K.S.Cole, Membranes, Ions and Impulses, University of California Press, Berkeley, (1968).

[67] B. Katz, Nerve, Muscle and Synapse, McGraw-Hill, New York, (1966).

[68] C. Steinem, A. Janshoff, J. Goossens, H-J. Galla, Bioelectrochem Bioenergetics, 45, (1998)17- 26.

[69] M. Khalid, G. Fouassier, H-J. Apell, F. Cornelius, and R.J. Clarke, Interaction of ATP with the Phosphoenzyme of the Na,K-ATPase, Biochemistry, 49, (2010) 1248-1258.

[70] M. Khalid, F. Cornelius, and R.J. Clarke, Dual mechanisms of allosteric acceleration of the Na+,K+-ATPase by ATP, Biophysial journal, 98, (2010), 2290-2298.

Chapter 2

Electron Beam Ablation Phenomenon – Theoretical Model and Applications

V.E. Ptitsin

Additional information is available at the end of the chapter

1. Introduction

The term "ablation" (i.e., removal or disposal of material) is quite often used in publications devoted to studies on physical and physicochemical processes caused by interactions of concentrated energy fluxes with absorbing condensed substance.

Most of the works on ablation relate to investigation or practical application of processes developed due to condensed matter interaction with high-power laser beams. In fact, now laser ablation is widely used in time-of-flight mass spectrometry as well as in modern micro- and nanoelectronics and microelectromechanics for structural surface modification and profiling of condensed substances.

However despite considerable academic and practical interest aroused by the ablation phenomenon, so far it has not been given a commonly adopted and (or) unique definition.

This may be so because of a too broad meaning put into the term ablation. Generally it means both the removal (disposal) of substance in the form of its certain fragments (droplets, macromolecules, clusters) under different physical conditions and fast phase transition of atoms and molecules of condensed matter from bound to free (excited or ionized) states. In such extended interpretation, this term seems to present no heuristic value.

In view of the aforesaid, in the present work the phenomenon of ablation is treated in a narrower physical sense, namely as a fast process of phase bound-free transition in excited or ionized structural elements (atoms, molecules) of condensed substance exposed to intense corpuscular radiation, e.g., intense electron fluxes.

It is known that interaction of a substance with intense either laser or electron beams exhibits some common regular features which have not yet been interpreted unambiguously. These are the following [1-5].

1. The phenomena of laser ablation and EBA are of threshold nature, i.e., they appear only when the beam intensity or the electron flux power density (P) exceeds some threshold value (P_l) characteristic of a given substance. For EBA in a wide range of condensed media, $P_l \approx 50$ MW/cm^2.
2. There is a finite time delay (Δt) between the onset of the substance exposure to a concentrated energy flux and the moment of phase transition of the substance to plasma. The time Δt strongly depends on the P level and can widely vary from $\approx 10^{-8}$ s ($P \geq P_l$) to $\approx 10^{-13}$ s (at $P \gg P_l$).
3. The charge content formed during dense plasma ablation also substantially depends on P. At the initial stage of plasma formation (at P close to P_l ($P > P_l$) native excited atoms (molecules) of a substance remain mostly neutral when going from bound to the free states. At high levels of P ($P \gg P_l$) the average charge number (Z) of plasma ions may be $Z \approx$ (2-3).

The purpose of the present part of the work is to study the physical mechanism of the above features. In the following sections of the paper we offer a phenomenological model of the EBA and interpretation of the two characteristic features of this phenomenon.

2. Theoretical model of the EBA phenomenon

2.1. Heat conducting as a limiting factor for the rate of energy absorption by a condensed substance

To calculate the rate of incident radiation energy absorption by a condensed substance (in units of W/m^2), we consider the condensed substance as a set of two interrelated subsystems: those of electrons and phonons.

It is known that when the substance surface is exposed to electromagnetic radiation or electron flux, the incident flux energy is mainly absorbed by the electron subsystem and then, due to electron-phonon interaction, is transferred to phonon subsystem. The energy transferred to the phonon subsystem propagates in the substance through the phonon-phonon interaction or the mechanism of heat conduction.

Based the above scheme of radiation energy absorption by the substance, we calculate estimates for the heat conduction flux power density (Θ) existing in this case. In a one-dimensional approximation, the heat flux propagating from the surface into the bulk of the substance, Θ can be expressed as [6]

$$\Theta \approx \frac{1}{3} \cdot n_f \cdot V_s \cdot \varepsilon, \tag{1}$$

where n_f is the phonon concentration, V_s is the sound velocity, ε is the average energy transferred by a phonon at its transition with a local temperature $T + \Delta T$ to the region with the temperature T. In Debye's approximation, the average phonon energy is defined by the known expression

$$\varepsilon \cong 9 \cdot \frac{N}{n_f} \cdot k\theta \cdot \xi^4 \cdot \int_0^{1/\xi} dx \cdot \frac{x^2}{e^x - 1}, \tag{2}$$

where N is the concentration of atoms in the substance, k is the Boltzmann constant, θ is the Debye temperature, $\xi \equiv T/\theta$. Integrating (2) in view of (1) yields

$$\Theta \cong Nk\theta V_s \xi \left[1 - \frac{3}{8\xi} + 3\sum_{n=1}^{\infty} \frac{(-1)^{n+1} B_n \xi^{-2n}}{(2n)! \cdot (2n+3)} \right], \tag{3}$$

where B_n are Bernoulli's numbers.

The analysis of the series in (3) shows its convergence, so if $\xi \geq 1$, Eq. (3) can be written in a more compact form:

$$\Theta \approx \gamma \cdot N \cdot k\theta \cdot V_s, \tag{4}$$

where $\gamma = \gamma(\xi)$ is a dimensionless parameter whose numerical value is close to γ ~0.5 at $\xi \approx$ 1. Eq. (4) shows that Θ only weakly depends on the substance nature since the respective substance characteristics present in (4) relatively little vary from substance to substance. Substituting the parameter values into (4) gives the value of Θ for a wide range of substances: $\Theta \approx (50 \pm 20)$ MW/cm^2.

Thus the obtained estimates of Θ and empirical values of P_l suggest that the existing ablation threshold results from the finiteness of the dissipation rate (in the phonon subsystem) of the radiation energy absorbed by the substance.

Besides, from the obtained estimate and the energy conservation law it follows that if the energy input velocity of the radiation incident on a condensed substance exceeds Θ, then the energy of the incident radiation (or electron flux) must be "reflected" by the substance in some way or another.

In the next section we obtain expressions for the ablation threshold P_l and delay time $\Delta t(P_l, P)$ by the example of interaction of an intense electron flux with metals. It is also shown that at $P > P_l$ the reflection of a concentrated energy flux by a substance may result from a fast phase transition of atoms (molecules) in the condensed substance from bound to free excited and ionized states. It seems evident that the physical "carrier" of the reflected energy flux during ablation is a flow of excited neutrals and ions of the condensed substance.

2.2. Modeling of the ablation process in the metal exposed to a concentrated energy flux – High-power density electron beam

To build the model of ablation, it would be natural to assume that the physical reason for the phase bound-free transition of surface excited or ionized atoms is that the binding (or cohesion) energy of atoms is reduced due to the excitation of the electron subsystem layer by external radiation or a high intensity corpuscular flux.

To build a model for the mechanism of processes occurring in a substance when the beam power density exceeds the level (Θ) which can be absorb through the mechanism of thermal conduction we consider the interaction of an intense electron flux with the metal surface. Note that such choice of the energy "carriers" and the "object" exposed to the energy flux is made to simplify description and analysis of the interaction of a concentrated energy flux with the substance.

The interaction between the electron and phonon subsystems is described based on the well known relaxation time approximation [3, 6].

Using the concepts developed in [6], the absolute value of binding energy at the excitation of the electron subsystem in the metal (Λ^*) can be defined as

$$\Lambda^* \approx \Lambda - r \cdot \left(\mu - \frac{3}{5} \cdot E_f \right) \tag{5}$$

where Λ is the absolute value of the surface atom binding energy in the absence of electron subsystem excitation, r ($r \approx 0.6 - 0.7$) is the dimensionless coefficient equal (in absence of electron subsystem excitation) to the average binding to cohesion energy ratio of the surface atom, μ is the average kinetic energy of the conducting electron, E_f is the Fermi energy. In the free electron model for metals the balance equation for the conducting electron kinetic energy μ in the metal exposed to an electron flux in a one-dimensional approximation can be written as [7-9]

$$\frac{d\mu}{dt} = \frac{(1-\eta) \cdot P \cdot \varphi(E)}{n \cdot E} - \frac{\varepsilon}{2\tau} \cdot \frac{k \cdot (T_e - T_p)}{E_f}, \tag{6}$$

where η is a dimensionless coefficient characterizing elastic electron scattering by the metal surface and also non-zero probability of emissive recombination of excited electrons (according to [3], $\eta < 0.1$), E is the energy of electron striking the metal surface, $\varphi(E)$ is the specific loss function of an electron injected into the metal subsurface layer; note that the $\varphi(E)$ value for the metal "surface" points should be calculated from the known Bethe formula for inelastic energy losses (see, e.g., [3]); besides, since Bethe's formula is approximate, in quantitative calculations, the theoretical value of this function needs correction based on the known experimental data; n is the conduction electron concentration in metal; P is the electron flux power density; τ is the characteristic time of electron–phonon interaction; T_p are the absolute electron and lattice temperatures, respectively.

Generally, integrating Eq. (6) requires specifying and finding the function $T_p \equiv T_p(t)$. However, as shown in [2], when the metal is exposed to intense electron fluxes, during the delay time (about $\approx 10^{-8}$ s) the lattice temperature increases only slightly and for, e.g., Cu it does not exceed ≈ 800 K. In this connection and taking into account that the heat capacity of a metal lattice is much greater than that of the electron gas, when integrating Eq. (5), one can assume $T_e >> T_p$ = const. The relationship between T_e and μ is of the form [10]

$$kT_e \approx \left[\mu^2 - \left(\frac{3}{5} \cdot E_f \right)^2 \right]^{1/2} \tag{7}$$

Using for Eq. (7) the approximation

$$kT_e \approx 3 \cdot \left(\mu - \frac{3}{5} E_f \right), \tag{8}$$

with a maximum error of 20% in the interval $0.6E_f \le \mu \le E_f$, from Eqs. (5-8) with the initial condition $\mu(0) = 0.6E_f$, we obtain

$$\Lambda^*(P,t) \approx \Lambda - \frac{2r \cdot (1-\eta) \cdot \tau E_f \varphi P}{3 \cdot \varepsilon \cdot n \cdot E} \times \left[1 - \exp\left(-\frac{3\varepsilon}{2E_f} \cdot \frac{t}{\tau} \right) \right] \tag{9}$$

Setting $\Lambda^*(P,t)$ equal to zero and assuming $t >> \tau$, from (9) one can find an expression for the minimum (or threshold) value of electron flux power density (P_l) which starts the excitation of the metal electron subsystem accompanied by the decrease in surface atom binding energy (down to zero). In other words, the conditions $\Lambda^*(P,t) = 0$, $t >> \tau$ physically mean that at $P \ge P_l$ after a certain time interval (delay time), "pumping" of the electron subsystem leads to a bound-free transition of excited and (or) ionized metal surface atoms not held by binding energies.

According to (9), the threshold level of power density can be expressed as

$$P_l \approx \frac{3 \cdot \varepsilon \cdot n \cdot E \cdot \Lambda}{2 \cdot r \cdot (1-\eta) \cdot \tau \cdot E_f \cdot \varphi} \tag{10}$$

Using the numerical estimates of from (10) for Cu, obtain

$$P_l \approx \Theta \tag{11}$$

The delay time Δt (at $P > P_l$) can be also calculated from Eq. (9). After some rearrangements from (9) we have

$$\Delta t \approx \frac{-2}{3} \cdot \frac{E_f}{\varepsilon} \cdot \tau \cdot \ln\left(1 - \frac{P_l}{P} \right) \tag{12}$$

The calculations of (12) show that in agreement with experimental data [1-4] Δt strongly depends on P and (for currently realized P levels) may vary from $\approx$ (10^{-7} - 10^{-8}) s to $\approx$ 10^{-13} s.

To conclude, we estimate the velocity of motion of the condensed substance-plasma boundary (v) in the course of ablation process. In the framework of the proposed model of ablation the activation energy value for the event of atomic bound-free transition can be assumed to be zero.

Based on the known concepts of Arrhenius–Frenkel about the probabilistic character of thermally activated bound-free transitions of surface atoms, the inter phase boundary motion velocity during the process of ablation can be expressed in the form of

$$v \approx \lim_{\Lambda^* \to 0} \left[af \cdot \exp\left(-\frac{\Lambda^*}{kT_p} \right) \right] = a \cdot \frac{kT_p}{h}, \tag{13}$$

where a is the lattice parameter,

$$f \approx \frac{kT_p}{h}$$

is the surface atom oscillation frequency. For example, for Cu one can easily access from (13) that already at a temperature T_p close to $\approx$ 500K, the velocity v reaches values $\approx V_s$. The estimates obtained for v and T_p are in quite satisfactory quantitative agreement with the data of [1, 2]. Eq. (13) also shows that during the process of ablation the ionic subsystem of a substance cannot heat substantially since the thermal front velocity in substance, in Debye's approximation does not exceed V_s. Besides, note that the proportionality between v and T_p and also the finiteness of the thermal front velocity must lead to mutual "self-consistency" of the velocities v and V_s and, hence, to their equalization during ablation.

Note that the inference made above about the relatively low temperature of the ion subsystem in the process of ablation does not contradict the known fact that ablation results in erosion markings and surface-scarred craters formed at the substance surface [1-4].

As consistent with the proposed ablation model, considerable heating and fusing of the substance may take place when $v < V_s$ and power density of the concentrated energy flux at the substance surface is less than P_l, accordingly. First of all such decrease in power density would always happen at the trailing edge of the energy pulse (down to 1 ns and less) high-power pulses are usually "bell-shaped". Second, if the energy pulse is relatively long, the level of the power density absorbed by the substance can fall to the values $P < P_l$ in a substantially shorter time than the pulse duration due to the growth of plasma thickness and density with time and, hence, of the efficiency of energy absorption in plasma.

Note, finally, that termination of the energy pulse action (irrespective of the pulse shape and width) and falling of P to $P < P_l$ would always be accompanied by the ion subsystem temperature rise over that in the ablation mode. This is because at $P < P_l$ the ionic subsystem of the subsurface layer will be heated as a result of both plasmoid disintegration and inertial absorption of the electron excitation energy by the ionic subsystem.

Thus, the EBA phenomenon is investigated on the basis of the general thermodynamic representations about power balance of absorption and dissipation energy processes in conditions of interaction of the concentrated energy flux with a condensed substance. It is shown that at high levels of absorbed power density, exceeding energy dissipation speed by the heat conductivity mechanism, the fast phase transition of a condensed substance in dense plasma can take place. The offered and developed general representations about the physical mechanism of the EBA phenomenon [11] allow suggest adequate interpretation (as qualitatively and quantitatively) of the basic laws of this phenomenon.

3. EBA phenomenon applications

In the recent two decades EBA phenomenon finds use in different areas of science and technologies [5, 9-20]. At the present time EBA is considered as one of alternatives to pulsed laser ablation (LA) which is more advanced. EBA possesses a number of important advantages: low-cost equipment (5-10% of the laser systems), big efficiency, possibility of ablation of optically transparent materials (e.g., SiO_2) and some others [12, 13]. EBA (or super-fast phase transition of solids into dense plasma plume) is one of the promising applications of high-intensive (more than 10 kA/cm²), low-energy (5-10 keV) pulsed (~100 ns) electron beams of small diameter (from ~ 1nm to ~10^6 nm and more).

By their physical nature the phenomena of LA and EBA are caused by the interaction of the concentrated energy fluxes with the condensed substance.

Of course the physical mechanisms of excitation and formation of plasma plumes for the processes LA and EBA have their special features; however, phenomenology of both processes, and in particular, the plasma density and also the dynamics of formation and dispersion of plasmoid (plume) are sufficiently close [5].

In connection with this the circle of scientific and technological tasks and problems, which can be solved on the basis of the use of the EBA phenomenon, to a considerable extent coincides with the spectrum of the tasks, solved at present by the LA method.

As it follows from the previous examination, EBA process is excited only when the electron beam power density P exceeds P_l

$$P = \left(\frac{1}{eS_b}\right) \cdot I_b \cdot E_b \geq P_l \cong 50\ MW/cm^2, \quad (14)$$

where e is the electron charge, S_b is the cross section of the beam on a substance (target) surface, I_b is the beam current, E_b (eV) is the beam energy on a substance (target) surface. Depending on the formulation of the scientific or technological problem the condition for the excitation EBA (14) can be realized by the electron beams, whose parameters: a radius (r_b) of beam, I_b, E_b can vary over wide limits.

Among the pioneer works on EBA study it is possible to note work [16]. For conducting of studies EBA phenomenon were used the pulsed (~ 100 ns) electron beams of small diameter (approximately of 1 mm) with these typical parameters: beam current density (J_b) - more than 10 kA/cm², beam energy 5-10 keV. In this work the dynamics of electron beam ablation plumes (SiO_2) have been characterized through the application of dye laser resonance absorption photography. Plume front expansion velocities of Si atoms were measured at nearly 1 cm/μs, and are comparable to the expansion of laser ablated metal atom plumes with laser fluences of a few J/cm².

For using the EBA phenomenon for the solution of the problems of micro- and the nano-technologies is, accordingly, necessary intensive ($P > 50\ MW/cm^2$) electron beams with a r_b from ~ 1 nm to ~ 1 mm. The formation of electron beams with such parameters is completely nontrivial task.

At the present time practical solution of this problem is obtained for the $r_b \approx 1$mm [21]. In the Neocera Company [21] are developed and in practice realized systems and installations for the formations of powerful pulsed electron beams (r_b ~ 1 mm), which permit implementation of a process of pulse electronic beam ablation. According to data of the Neocera Company process EBA is the process in which a pulsed (100 ns) high power electron beam (approximately 10^3 A, 15 keV) penetrates approximately 1 μm into the target resulting in a rapid evaporation of target material, and its transformation in plasma state. The non-equilibrium extraction of the target material (ablation) facilitates stochiometric composition of the plasma. Under optimum conditions, the target stoichiometry is thus preserved in the deposited films. All solid state materials-metals, semiconductors and insulators, can be deposited as thin films with EBA. As in the case of Pulsed Laser Deposition (PLD), the pulsed electron deposition (PED) technique provides a unique platform for depositing thin films of complex materials on a variety of technologically important substrates, with a unique strength of extending the range of materials and applications. Unlike PLD, where the ablation process is critically dependent on the optical absorption coefficient of the target material, in PED, the ablation depends only on the range of electrons in the target. For most of the solid state materials, this range is of the order of a few microns. For example, SiO_2 with a large optical band-gap of 10eV is transparent to the 248 nm Kr-F excimer laser radiation. In PED technique however, the high-power electrons can strongly couple to the target material (SiO_2), leading to SiO_2 film deposition. The beam-solid interaction mechanism is quite different in PED in comparison to PLD. This unique difference provides thin film experimentalists a mechanism to extend the parameter-space required for certain novel materials fabrication. Let us note also some representative materials systems explored by PED [21]:

- High-Temperature Superconducting (HTS) YBCO (and GdBCO) films.
- Metallic-oxide (SrRuO3) films.
- Paraelectric (Ba-SrTiO3) films.
- Insulating glass (SiO2-based) films and Al2O3 films.

However, formation of high intensity electron beams in the so-called sub-micron range r_b from ~0.01 μm to ~ 1 μm up to now are not thus far created. In this direction thus far there are only separate studies [14, 15], in which is theoretically and experimentally substantiated the possibility of designing of beams with the high power density (up to ~100 GW/cm^2).

Thus, for the range of $0.01\mu m \le r_b \le 1\mu m$ the problem of the creation forming systems of the electron beams, the power density of which would satisfy condition (14) as long as remains not solved.

Considerable achievements in the region of the practical applications of the EBA phenomenon in science the nanotechnology it was possible to attain as an instrument the transmission electron microscopy (TEM) technique [17 - 20].

This based on the EBA phenomenon method [17 - 20] of the studies is used as the investigation tool a high-resolution transmission electron microscope (TEM) to fabricate various (with sub-nm resolution) nanostructures and nanodevices including nanogaps, nanodiscs, nanorings, nanochannels, nanowires with tailored curvatures and multi-terminal nanogap devices with nanoislands or nanoholes between the terminals. This method was called TEBAL (transmission electron beam ablation lithography). During TEBAL, the user is able *in situ* to watch on the fluorescent imaging screen of the TEM the ablation process of the unwanted metal, followed by the crystallization of the surrounding metal and the formation (Fig.1) of desired geometries [17]. The high resolution, geometrical flexibility and yield make this fabrication method unique and attractive for many applications including nanoelectronics, nanofluidics, plasmonics, nanoparticle/atom manipulation on a chip. Particularly exciting and unique are the nanogap-nanohole devices as they open new opportunities for DNA nanopore sequencing [22].

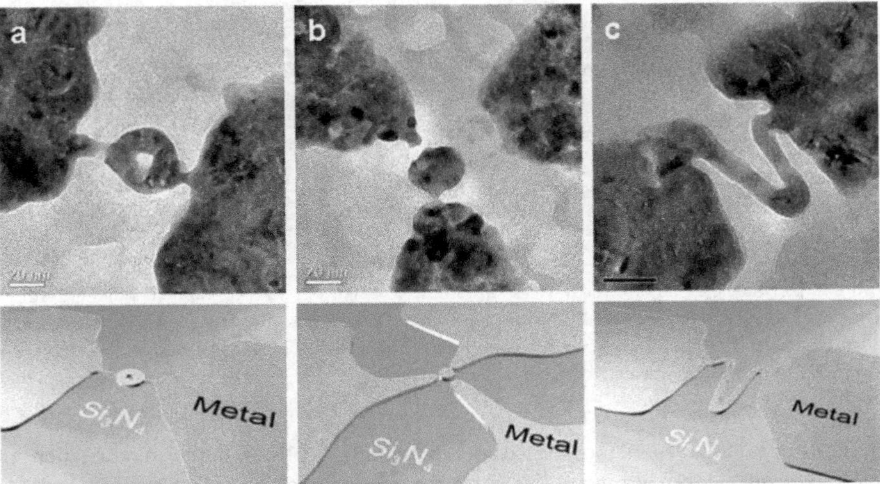

Figure 1. [17]. Example structures to demonstrate the flexibility of TEBAL. Each of the three structures shown in the TEM images is accompanied by a schematic (below) showing the fabrication by TEBAL. (a) Nanoring with outer radius of ~18.5 nm and inner radius of ~3 nm (scale = 20 nm). (b) Three-terminal electronic device: source and drain leads are coupled to a ~13 nm radius metallic island and a gate electrode ~23 nm away from the island (scale =20 nm). The rate-limiting tunneling barrier (upper junction) is a 2.7 nm gap. (c) Serpentine wire with 6 nm width (scale = 20 nm). All lengths were measured with Gatan's Digital Micrograph image analysis software.

In the recent work [20] it communicates about the electronic measurements on high quality graphene nanoconstrictions (GNCs) fabricated in a transmission electron microscope (TEM), and the first measurements on GNC conductance with an accurate measurement of constriction width down to 1 nm. To create the GNCs, freely suspended graphene nanoribbons (GNR) were fabricated using few-layer graphene grown by chemical vapor deposition. The ribbons were loaded into the TEM, and a current-annealing procedure was used to clean the material and improve its electronic characteristics. The TEM beam was then used to sculpt GNCs to a series of desired widths in the range 1-700 nm; after each sculpting step, the sample was imaged by TEM and its electronic properties were measured in situ. The diagram of experiments [20] and some results are given in figure 2. GNC conductance was found to be remarkably high, comparable to that of exfoliated graphene samples of similar size. The GNC conductance varied with width approximately as $G(w) = (e^2/h)\ w^{0.75}$, where w is the constriction width in nanometers. GNCs support current densities greater than ~ 120 $\mu A/nm^2$, 2 orders of magnitude higher than that which has been previously reported for graphene nanoribbons and 2000 times higher than that reported for copper.

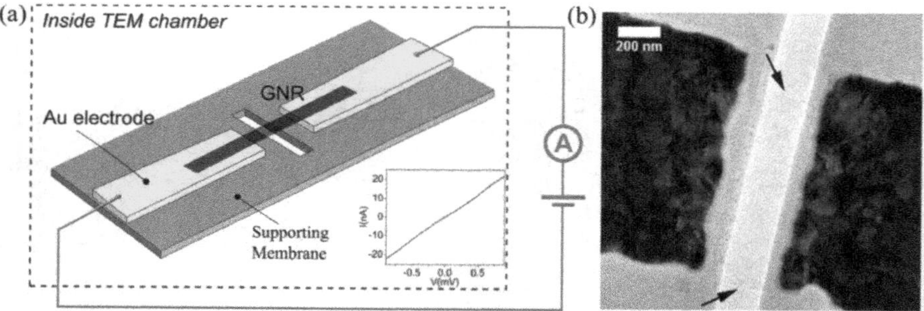

Figure 2. [20]. Suspended graphene devices. (a) Sample schematic. Few layer graphene ribbon (3 -10 layers thick) is suspended over a 1.4 μm x0.2 μm slit in a 100 nm thick silicon nitride (SiN) membrane (membrane size ~40 μm x 40 μm). Inset: current-voltage characteristic of an as-fabricated nanoribbon, acquired in situ. (b) TEM image of a suspended graphene nanoribbon. Arrows indicate the edges of the graphene.

In the work [18] was demonstrated the possibility high-resolution modification of suspended multi-layer graphene sheets by controlled exposure to the focused electron beam of a transmission electron microscope. It was shown that this technique can be used to realize, on timescales of a few seconds, a variety of features, including nanometer-scale pores, slits, and gaps that are stable and do not evolve over time. Despite the extreme thinness of the suspended graphene sheets, extensive removal of material to produce the desired feature geometries is found to not introduce long-range distortion of the suspended sheet structure. Some results of these impressive experiments are given in figures 3 and 4.

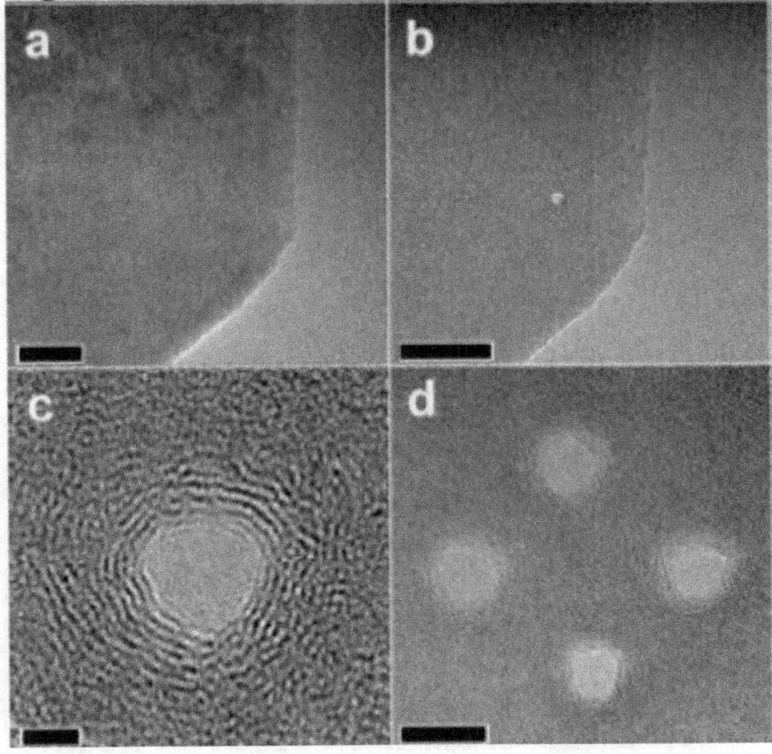

Figure 3. 18]. TEM images of a suspended graphene sheet is shown (a) before and (b) after a nanopore is made by electron beam ablation. (c) Higher magnification image of the nanopore. (d) Multiple nanopores made in close proximity to each other. (Scale bars are 50, 50, 2, 10 nm).

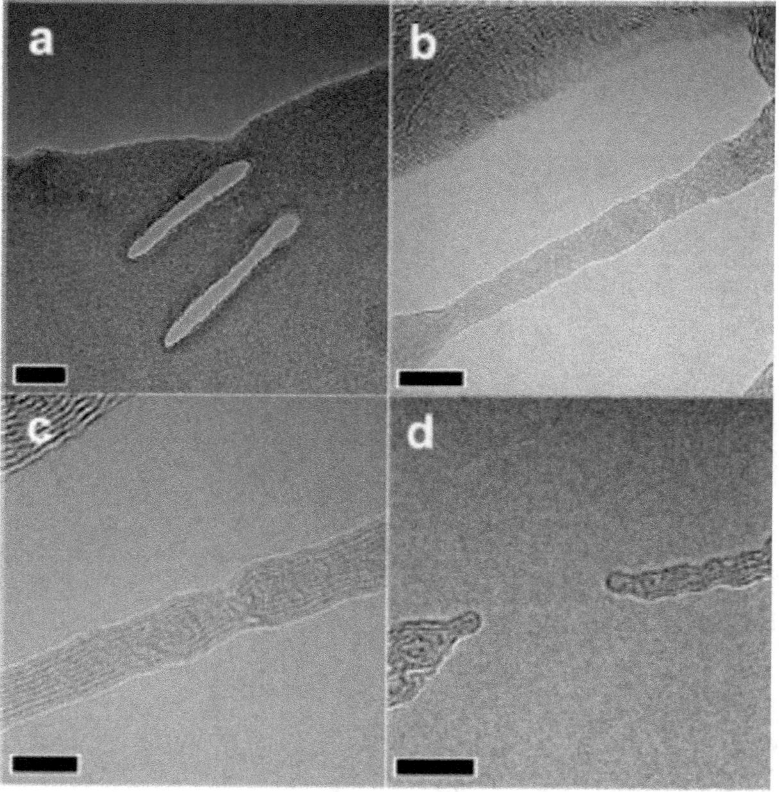

Figure 4. [18]. (a) Two ~ 6 nm lines cut into a graphene sheet. (b) Electron irradiation is continued to create a ~ 5nm wide bridge. (c) Higher resolution of the bridge shows clear atomic order. (d) Small gap opened in the nanobridge by additional electron irradiation. Note that the cut ends are closed. (Scale bars are 20, 10, 5, 5 nm).

Author details

V.E. Ptitsin

Institute for Analytical Instrumentation of the Russian Academy of Sciences, Saint – Petersburg, Russia

Acknowledgement

The work has been supported by the Russian Academy of Sciences and the Russian Foundation for Basic Research.

4. References

[1] J. Ready, Effect of High-Power Laser Radiation, Ed. S. I. Anisimov (Mir, Moscow, 1974) 468 p.

[2] G. A. Mesyts and D. I. Proskurovsky, Pulsed Electric Dicharge in Vacuum (Springer, Heidelberg, 1989) 256 p.

[3] K.A. Valiev, Physics of the Submicron Lithography (Nauka, Moscow, 1990) 528 p.

[4] Yu. A. Bykovsky, V. N. Nevolin, Laser Mass Spectrometry (Energoatomizdat, Moscow, 1985) 128 p.

[5] S. D. Kovaleski, R. M. Gilgenbach, L. K. Ang, Y. Y. Lau and J. S. Lash, Appl. Surf. Sci. 127–129, 947 (1998).

[6] J. Ziman, Electrons and Phonons (Mir, Moscow, 1962) 354 p.

[7] M. I. Kaganov, I. M. Lifchits and L.V. Tanatarov, ZhETF, 232 (1956).

[8] V. E. Ptitsin, Doctorate Thesis (Phys. &Math.), IaI RAS, (1996).

[9] V. E. Ptitsin, Rev. Sci. Instr., 65, No. 4, 1476 (1994).

[10] Yu. V. Martynenko and Yu. N. Yavlinsky, Moscow Nuclear Power Institute Preprint, No. 4084/11, Moscow, (1985).

[11] V. E. Ptitsin, Phys. Status Solidi C 9, No. 1, 15–18 (2012)

[12] G. Muller, C. Schulteiss, Proc. 10th Int. Conf. on High Power Particle Beams (BEAMS-94), 1994. pp. 833-836.

[13] W. Frey, C. Schulteiss, H. Bluhm, Proc. 14th Int.Conf. on High Power Particle Beams (BEAMS-2002), 2002. pp. 87-90.

[14] V.E. Ptitsin, V.F. Tregubov, Nauchnoije Priborostroenije, (Science, Saint- Petersburg), vol. 12, No 3, pp. 15 – 25, (2002)

[15] V.E. Ptitsin, V.F. Tregubov, Ultramicroscopy, v. 95, pp. 131-138 (2003).

[16] S. D. Kovaleski, R. M. Gilgenbach, L. K. Ang, and Y. Y. Lau, Appl. Phys. Lett., vol. 73, No. 18, pp. 2576-2578 (1998).

[17] Michael D. Fischbein and Marija Drndic, Sub-10 nm Device Fabrication in a Transmission Electron Microscope, Nano Letters 7 (5), 1329-1337, (2007).

[18] Michael D. Fischbein and Marija Drndic, Electron beam nanosculpting of suspended graphene sheets, Applied Physics Letters 93, 113107, (2008).

[19] T. Gnanavel, Z. Saghi, M A Mat. Yajid, Y. Peng, B. J. Inkson, M. R .J. Gibbs and G. Mobus, Journal of Physics: Conf. Ser. 241 (2010) 012075.

[20] Ye Lu, Christopher A. Merchant, Marija Drndic, A.T. Charlie Johnson, In Situ Electronic Characterization of Graphene Nanoconstrictions Fabricated in a Transmission Electron Microscope, Nano Letters, 11 (12) 5184-5188, (2011).

[21] Neocera Company; see: www.neocera.com

[22] Christopher A. Merchant, Ken Healy, Meni Wanunu, Vishva Ray, Neil Peterman, John Bartel, Michael D. Fischbein, Kimberly Venta, Zhengtang Luo, A. T. Charlie Johnson, Marija Drndic, DNA translocation through graphene nanopores, Nano Letters, 10 (8), pp 2915-2921, 2010.

Chapter 3

Oxidation Chemistry of Metal(II) Salen-Type Complexes

Yuichi Shimazaki

Additional information is available at the end of the chapter

1. Introduction

The oxidation chemistry of metal complexes has been widely developed in recent years, affording deep insights into the reaction mechanisms for many useful homogeneous catalytic reactions and enzymatic reactions at the active site of metalloenzymes [1]. In the course of the studies, a large number of novel complexes have been synthesized and well characterized [2-18], and especially high valent metal complexes formed as a result of redox reactions have become important in catalytic and biological systems. In general high valent metal complexes have been meant to show the complexes oxidized at the metal center, and the formal oxidation state is identical with the oxidation state of the central metal ion reported in [19-23]. For example, one-electron oxidation of potassium hexacyanoferrate(II), $K_4[Fe(CN)_6]$, gives potassium hexacyanoferrate(III), $K_3[Fe(CN)_6]$, whose valence state of the central iron ion is +III and agrees with the experimental valence state of the ion. In contrast to this, the formal oxidation number of the central metal ion of the complexes of iminophenolate dianion, $(L^{AP})^{2-}$ is not always identical with the experimental valence state reported in references [24,25]. In the case of $[Ni(L^{AP})_2]^0$, the formal oxidation state of the central nickel ion is +IV, but the experimental valence state of nickel can be assigned to be +II, and two iminophenolate dianions are oxidized to iminosemiquinonate radical anions $(L^{SQ})^-$ (**Figure 1**).

Such a difference between the formal oxidation number and the experimental oxidation state is also observed in biological systems. Recently, it has been reported that various radicals can be generated at a proximal position of the metal center in metalloproteins, and the radical can sometimes interact with the central metal ion as shown in reference [21]. Galactose oxidase (GO) is a single copper oxidase, which catalyzes a two-electron oxidation of a primary alcohol to the corresponding aldehyde [21-24]. The active site structure of the inactive form of GO is shown in **Figure 2**, where two imidazole rings of histidine residues,

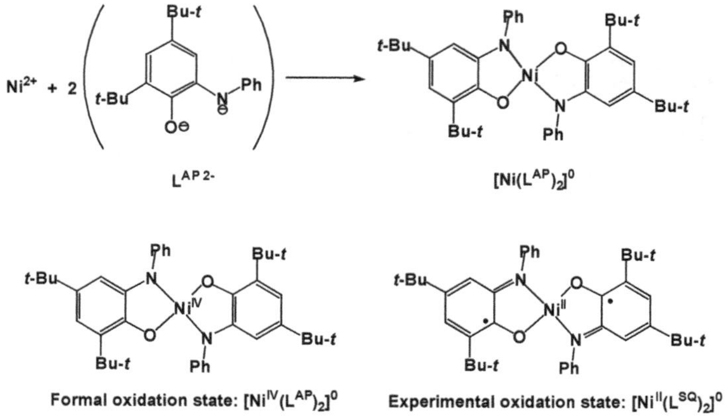

Ni^{2+} + 2 $L^{AP\,2-}$ → $[Ni(L^{AP})_2]^0$

Formal oxidation state: $[Ni^{IV}(L^{AP})_2]^0$ Experimental oxidation state: $[Ni^{II}(L^{SQ})_2]^0$

Figure 1. An example showing the difference between the experimental oxidation state and the formal oxidation number in the Ni complexes of iminophenolate dianions, $[Ni(L^{AP})_2]^0$.

two phenol moieties of tyrosine residues, and an acetate ion are coordinated to the copper(II) ion [24]. One of two phenol moieties is in the deprotonated form and is coordinated to the copper ion at an equatorial position, and another phenol moiety is protonated and located at an apical position.

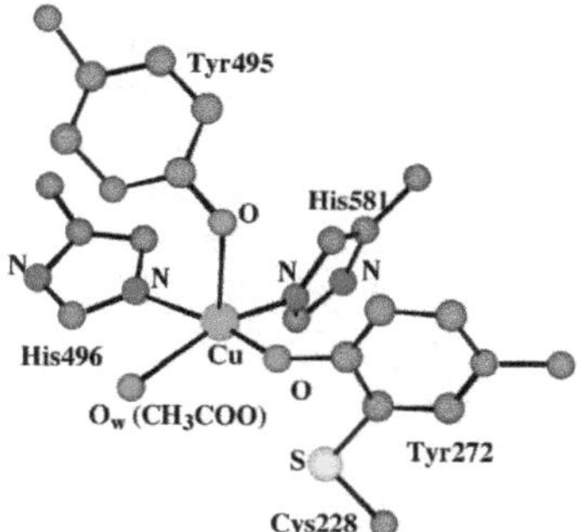

Figure 2. Active site structure of the inactive form of GO.

Conversion to the active form of GO occurs upon one-electron oxidation and deprotonation from the apical phenol moiety, which gives the structure with the two phenolate moieties coordinated to the copper center. This active form should act as a two-electron oxidant, causing the conversion from primary alcohol to aldehyde. Therefore, the formal oxidation state of the active form of GO can be described as a Cu(III)-phenolate species. Actually, the active form of GO had been considered to be a copper(III) species [25,26], but early in 1990's, various spectroscopic studies of the active form of GO revealed the formation of the phenoxyl radical species and the Cu(II)–phenoxyl radical bond. The free phenoxyl radical is very unstable with half life estimated to be 2.4 ms at ambient conditions, while the Cu(II)–

phenoxyl radical in the active form of GO has a long life; the radical is not quenched for more than one week at room temperature [22,23,27,28]. Thus, properties of the metal coordinated phenoxyl radical show a significant change from those of the free phenoxyl radical.

The proposed reaction mechanism of GO is that the primary alcohol is coordinated to the Cu(II)-phenoxyl radical species generated by molecular oxygen and oxidized by an intramolecular two-electron redox reaction with the hydrogen atom scission from the alcohol moiety to give the aldehyde and the Cu(I)–phenol species (**Scheme 1**) reported in [22,23]. The Cu(I)-phenol complex is oxidized by molecular oxygen to regenerate the Cu(II)-phenoxyl radical species.

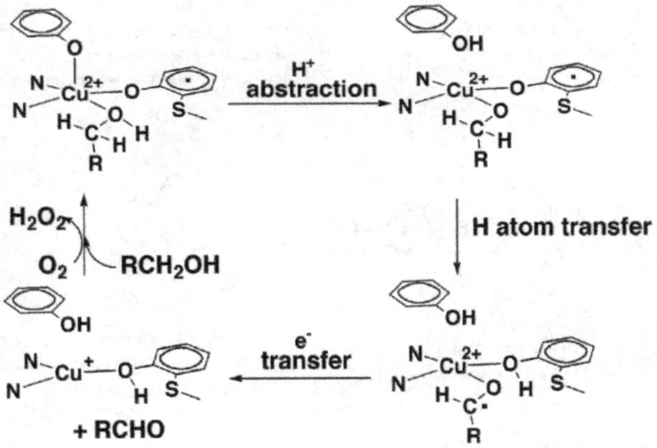

Scheme 1. Proposed mechanism of galactose oxidase (GO).

For understanding the detailed mechanism of GO and the properties of the metal complexes with the coordinated phenoxyl radical, many metal–phenolate complexes have been synthesized, and their oxidation behavior and properties of the oxidized forms have been characterized in [12,13, 29-36]. Salen (Salen = di(salicylidene)ethylenediamine) and its family are one of the most important and famous ligands having two phenol moieties (**Figure 3**) [37,38]. One of the characteristics of Salen is its preference for a square planar 2N2O coordination environment, while the distortion from the square plane can be introduced by changing the diamine backbone. A number of metal-Salen complexes have been reported to be the very important catalysts for oxidation and conversion of various organic substrates from early 1990's [37]. While studies on oxidative reaction intermediates are in progress, the oxidation state of the metal ions in the active species has not been fully understood until now. Detailed descriptions of the oxidation state of the intermediate are sometimes complicated, because the oxidation locus on oxidized metal complexes is often different from the "formal" oxidation site [19,20]. Although "formal" and "experimental" oxidation numbers are identical in many cases, they are often used as synonyms, since the term of the physical or experimental oxidation state has not been accepted in some areas of chemistry.

The present argument is focused on recent advances in the chemistry related with the synthesis, characterization, and reactivity of some of the one-electron oxidized metal(II)–salen type complexes, especially the complexes of group 10 metal(II) ions, Ni(II), Pd(II), and Pt(II) [39-43], and Cu(II) complexes [44-47] (**Figure 3**).

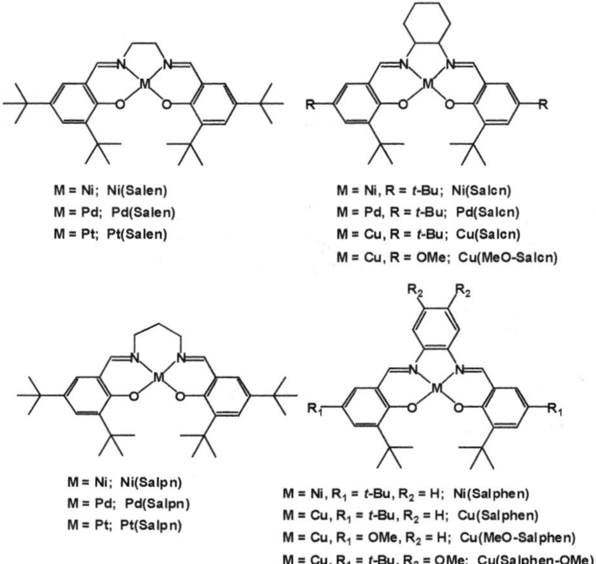

Figure 3. Abbreviations of salen-type complexes.

Various useful and interesting salen-type complexes have been synthesized and characterized until now [48], and detailed electronic structural studies have recently been reported for a dinuclear chelating salen-type ligand having a catechol and a tetra(amino)tri(hydroxy)phenyl moiety [49, 50]. In view of the interest in the redox properties of metal-phenolate complexes, detailed electronic structures of the one-electron oxidized complexes are discussed in order to understand the electronic structure difference between the metal-centered and ligand-centered oxidation products and its dependence on the properties of metal ions and substituents of the phenolate moiety. For this reason, this argument is described only for a few examples whose detailed electronic structures have been clarified.

2. Redox potentials of phenol and metal(II) salen-type complexes

One-electron oxidation of closed-shell organic molecules is generally difficult, and the one-electron oxidized products may be unstable. Oxidation of the free phenol similarly gives the unstable phenoxyl radical described in the introduction and references [21-23,27,28]. The potential of formation of the phenoxyl radical is predicted to be high, and actually the potential of a tri(*tert*-butyl)phenol was estimated to be +1.07 V as reported in [51]. On the other hand, the oxidation potential of the phenolate anion is much lower (-0.68 V) in comparison with that of phenol [52], suggesting that deprotonation from phenol is favorable for formation of the free radical .

The oxidation chemistry of metal phenolate complexes has shown that the oxidation potentials of phenolate complexes are intermediate between those of free phenolate and free phenol [53]. Such a trend is also applied to the salen-type complexes, and most of the Cu(II) and group 10 metal(II) salen-type complexes exhibited two reversible redox waves in the range of 0 to 1.5 V vs NHE, due to having two phenolate moieties [29-31, 39-47]. On the other hand, the metal centered one- and two-electron oxidized complexes may possibly be generated in the similar potential range. Some copper(III) [12,13] and nickel(III) [54,55] complexes have been reported, and two-electron oxidized species of the group 10 metal ions, especially some Pd(IV) and Pt(IV) complexes, have been reported. Some of the metal complexes such as K_2PdCl_6 and K_2PtCl_6 are commercially available [56]. Therefore, the experimental oxidation state can not be determined from the oxidation potential only. The redox potentials of some salen-type complexes are listed in **Table 1,** and the voltammograms of two copper(II) complexes are shown as examples in **Figure 4** [39-47].

complex	potential (E / V vs. Fc / Fc$^+$)	complex	potential (E / V vs. Fc / Fc$^+$)
Ni(Salcn)	0.46, 0.80	Ni(Salphen)	0.58, 0.80
Pd(Salcn)	0.45, 0.80	Cu(Salcn)	0.45, 0.65
Pt(Salen)	0.35, 0.94	Cu(MeO-Salcn)	0.28, 0.44
Ni(Salpn)	0.43, 0.69	Cu(Salphen)	0.65, 0.83
Pd(Salpn)	0.52, 088	Cu(MeO-Salphen)	0.38, 0.49
Pt(Salpn)	0.44, 0.99	Cu(Salphen-OMe)	0.41, 0.70

Table 1. Redox potential of Cu(II) and group 10 metal complexes

From the list of **Table 1**, the ranges of both first and second redox potentials are relatively narrow, especially the range of the first redox potentials being from 0.28 to 0.65 V. Such a narrow range may be generally ascribed to the fact that the all the complexes have a similar oxidation locus. The small potential differences are due to the substitution of the phenolate moiety and the ligand structure. However, there are some different electronic structures among the one-electron oxidized forms of the complexes in Table 1, i.e., a M^{III}-phenolate ground state complex, a M^{II}-phenoxyl radical where the radical electron is fully delocalized on two phenolate moieties, a relatively localized M^{II}(phenolate)(phenoxyl) complex, and a M^{II}(dinitrogen ligand radical)(phenolate)$_2$ species [39-47]. The similarity of the potentials indicates that formation of all of the oxidized species is due to a simple electron transfer without significant structural changes. Thus, the experimental oxidation state of the oxidized complexes cannot be determined only from the redox potentials.

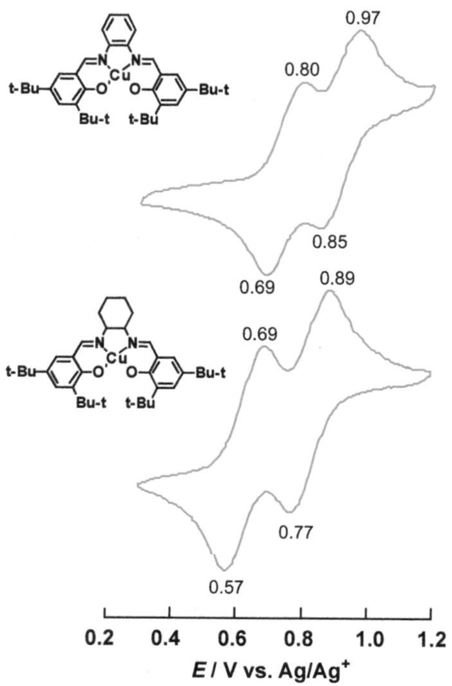

Figure 4. Cyclic voltammograms of Cu(II) complexes: Top, Cu(Salphen); bottom, Cu(Salcn).

3. Isolation and solid state characterization of one-electron oxidized complexes

One-electron oxidized complexes have been synthesized by reaction with a one-electron oxidant, such as Ce^{IV}, Ag^{I}, NO^{+}, some organic reagents, and so on [57]. For most complexes shown in **Figure 3**, one-electron oxidized species were generated by addition of $AgSbF_6$ to the CH_2Cl_2 solution of metal(II) salen-type complexes. Ag^{I} with a potential of 0.799 V vs. NHE can act as a one-electron oxidant for complexes [57,58]. The oxidation method using Ag ion is useful for generation of the relatively stable oxidized complexes, since Ag^{0} is an easily removable product formed in the course of the oxidation. Some solutions of one-electron oxidized complexes were kept standing for a few days to give the products as crystals.

The X-ray crystal structure analyses of one-electron oxidized group 10 metal salen-type complexes are shown in **Figure 5** [41-43]. The structures of all these complexes were found to be similar to those of the corresponding complexes before oxidation, which supports the CV results that significant structure changes did not occur in the course of the oxidation.

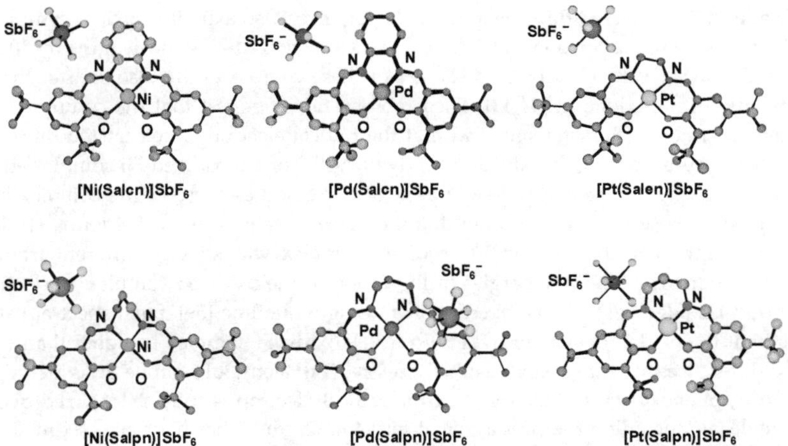

Figure 5. Crystal structures of one-electron oxidized group 10 metal salen-type complexes

However, a close look into the details of the crystal structures reveals that there are subtle differences between them, and especially the oxidized Pd(II) complexes are different from the other complexes [42,43]. Comparison of the 5-membered dinitrogen chelate backbones of the Salcn and Salen complexes indicates that upon oxidation the Ni and Pt complexes exhibited a clear coordination sphere contraction due to shortening of the M–O and M–N bond lengths. On the other hand, the Pd complex showed an unsymmetrical contraction [42]: One of the Pd–O bonds (2.003 Å) is longer than the other (1.963 Å), and the C–O bond (1.263 Å) of the phenolate moiety with the longer Pd–O bond is shorter than the other C–O bond (1.317 Å). The phenolate moiety with a shorter C–O bond length has the lengthening of the ring *ortho* C–C bonds in comparison with those of the other one. These structural features of the phenolate moiety in the oxidized Pd complex are in good agreement with the characteristics of the phenoxyl radical, which showed the quinoid form due to delocalization of the radical electron on the phenolate moiety as shown in **Scheme 2** and reference [53]. Such properties were also detected for the Pd complex with the 6-membered dinitrogen chelate back bone, [Pd(Salpn)]SbF_6 [43]. In addition, the SbF_6^- counterion was positioned close to the quinoid moiety of this complex; the closest distance between the SbF_6^- and the C-O carbon atom of the phenoxyl ligand was 3.026 Å. Therefore, one-electron oxidized Pd(II) complexes can be assigned to relatively localized Pd^{II}(phenoxyl)(phenolate) complexes.

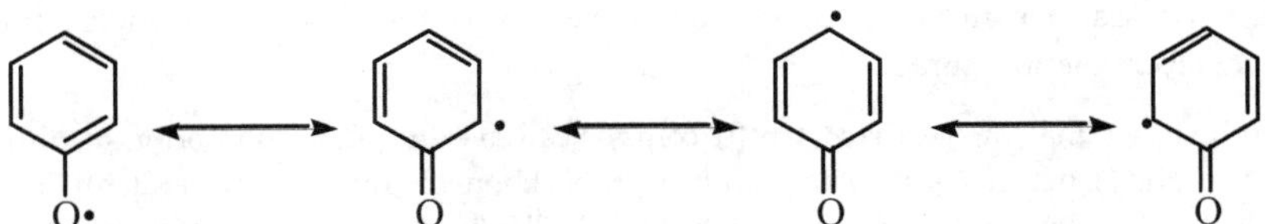

Scheme 2. Canonical forms of the phenoxyl radical

The Ni and Pt 5-membered dinitrogen chelate complexes also exhibited a clear symmetrical coordination sphere contraction in both two M–O and two M–N bond lengths (ca. 0.02 Å) upon oxidation, and the C–O bond distances of these complexes are also shorter than the same bonds before oxidation [41,43]. These observations suggest that the complexes have the phenoxyl radical characteristics and that the radical electron is delocalized on the two phenolate moieties. Indeed, the XPS and K-edge XANES of an oxidized Ni complex showed the same binding energies and pre-edge peak of nickel ion as those of the complex before oxidation [40]. These results supported that the valence state of the nickel ion is +II. In the case of Pt complexes, the XPS of the oxidized complex was slightly different from that before oxidation. The binding energies of the Pt ion in the oxidized complex were +0.2 eV higher, and L_{III}-edge XANES exhibited an increasing white line [59]. Such spectral features suggest that the oxidation state of the Pt ion in the oxidized complex is higher than +II but that the differences are rather small [40]. Therefore, [Pt(Salen)]SbF_6 can be described mainly as the Pt(II)-phenoxyl radical species, but the radical electron is fully delocalized over the whole molecule including the central metal ion [40]. On the other hand, the six-membered Ni^{II} and Pt^{II} Salpn chelate complexes are slightly different from the 5-membered dinitrogen chelate Salcn and Salen complexes [43]. Crystal structures of both oxidized Salpn complexes exhibited two crystallographically independent molecules in the unit cells. The M–O and M–N bond lengths do not differ substantially between the two molecules in the unit cell. The bond lengths in the coordination plane are ca. 0.02 Å shorter than those of the neutral complexes, and this contraction upon oxidation is in good agreement with the 5-membered dinitrogen chelate complexes. However, the C–O bond lengths of the two phenolate moieties differed for the two independent molecules; one of the molecules showed very similar C–O bond lengths, while the bond lengths in the other molecule were slightly different, showing a similar tendency to that of the oxidized Pd complexes. Therefore, the 6-membered Ni and Pt chelate complexes can be considered to be closer to the localized phenoxyl radical metal(II) complexes in comparison with the 5-membered chelate complex, Salcn and Salpn due to the chelate effect of the dinitrogen backbone [43]. However, determination of the detailed electronic structure of oxidized salen-type complexes, especially 5-membered Ni and Pt chelate complexes, only from the X-ray crystal structure analysis may be difficult, since we can detect mainly the contraction of the coordination sphere, which is predicted to be also observed in high valent metal salen-type complexes.

On the other hand, the electronic structure determination of the Cu complexes are clearly made by X-ray structure analyses. Structures of three one-electron oxidized Cu^{II} salen-type complexes are shown in **Figure 6** [44,45,47]. X-ray analyses of all these complexes established that their structures are similar to those before oxidation, indicating a simple one-electron transfer from these precursors. However, the three Cu complexes have different electronic structures.

Figure 6 shows that the oxidized Cu(II) complexes have the SbF_6^- counterion at different positions. The structures of the same dinitrogen backbone complexes, [Cu(Salcn)]SbF_6 and [Cu(MeO-Salcn)]SbF_6, indicate that a weak axial Cu-F interaction (2.76 Å) exists between the counterion and the metal center in $[Cu(Salcn)]^+$ [45], whereas $[Cu(MeO\text{-}Salcn)]^+$ has a weak

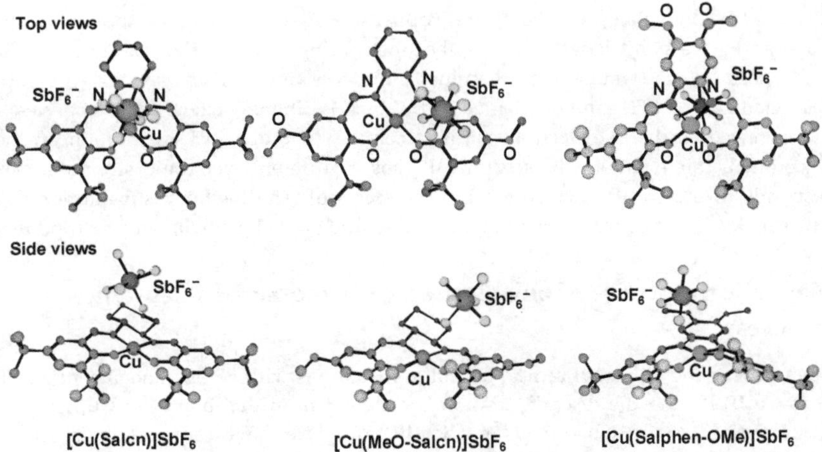

Figure 6. Crystal structures of one-electron oxidized Cu(II) salen-type complexes

F–C interaction between the counterion and one side of the phenolate moieties [44]. This difference suggests that the oxidation locus of these complexes are different; [Cu(Salcn)]SbF_6 has a Cu(III) character, while [Cu(MeO-Salcn)]SbF_6 is a Cu(II)-phenoxyl radical complex. [Cu(Salcn)]SbF_6 showed contraction of the coordination sphere without shortening of the C–O bonds of both phenolate moieties and distortion of the coordination plane was substantially reduced from that in Cu(Salcn) [45]. Such structural features are in good agreement with those of the low-spin d^8 Cu(III) complexes [12,13]. Indeed, the XAF and XPS studies of [Cu(Salcn)]SbF_6 reported the Cu(III)-phenolate ground state, because the pre-edge of oxidized complex was more than 1 eV higher, and the binding energies of Cu ion in the oxidized complex was also 1 eV higher as compared with the neutral complex Cu(Salcn) [45]. These characteristics are in good agreement with the Cu(III) valence state [12,13]. On the other hand, the MeO-substituted complex [Cu(MeO-Salcn)]SbF_6 exhibits that the C-O bond of one-side of the phenolate moieties is shortened and that the Cu–O bond with the shortened phenolate moiety becomes longer [44]. The counterion is close to the phenolate moiety of the shortened C-O bond. Such characteristics are in good agreement with those of the oxidized Pd(Salpn) having a relatively localized Pd^{II}(phenoxyl)(phenolate) structure. Therefore, [Cu(MeO-Salcn)]SbF_6 can be described as a localized Cu^{II}(phenoxyl)(phenolate) complex [44].

Another one-electron oxidized complex, [Cu(Salphen-OMe)]SbF_6, has a different electronic structure from the previous two Salcn complexes [47]. The C–C and C–O bond lengths within the phenolate rings do not differ significantly from those of the complex before oxidation, Cu(Salphen-OMe). Although there is only a slight contraction of the Cu-O and Cu-N bonds, the copper ion geometry is significantly distorted toward a tetrahedral geometry in our case (the dihedral angle of 22° between the O1-Cu-N1 and O2-Cu-N2 planes). These structural features indicate that the oxidation locus of [Cu(Salphen-OMe)]$^+$ is neither the two phenolates nor the central copper ion. A striking feature upon oxidation is the change in the bond lengths within the phenyl ring. Further the counterion SbF_6^- is located close to the *o*-phenylenediamine moiety. Therefore, the oxidized complex, [Cu(Salphen-OMe)]SbF_6, can be assigned to the Cu(II)-diiminobenzene radical species [47]. In this connection, the one-electron oxidized complex without any substitution on the *o*-phenylenmediamine moiety, $[Cu(Salphen)]^+$, has a different electronic structure, which corresponds to the Cu(II)-phenoxyl radical species [46]. The methoxy substitution in the phenyl ring leads to a different electronic structure due to its electron donating property.

4. Magnetic properties of one-electron oxidized metal salen-type complexes

The behavior of the phenoxyl radical bound to metal ions with an open-shell configuration such as Cu(II) having the d^9 configuration is different from that bound to Cu(III) ground state complex, [Cu(Salcn)]SbF_6. In general, Cu(III) complexes have a square-planar geometry and are diamagnetic and EPR silent due to the low–spin d^8 configuration [12,13]. The Cu(II)–phenoxyl radical complex, on the other hand, has two electron spins on different nuclei, and therefore the spin–spin interaction should be considered in [29-31]. In the case of the oxidized Cu(II) salen-type complexes, since correlation between ligand *p*-orbital and the copper $d_{x2\text{-}y2}$ orbital having an unpaired electron is close to orthogonal, the *d*-electron spin of copper ion coupled with radical electron spin ferromagnetically [60].

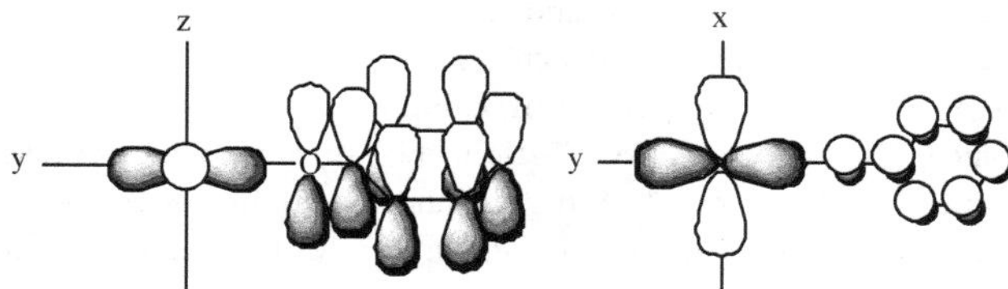

Figure 7. Schematic view of the orthogonality between ligand *p*-orbital and the copper $d_{x2\text{-}y2}$ orbital

Magnetic properties of [Cu(Salcn)]SbF_6 in the solid state have been reported to be a temperature independent diamagnetic species with the effective magnetic moment m_{eff} = 0.3 $m_{B.M.}$ at 300 K, which is in good agreement with the considerations from the crystal structure analysis[25]. On the other hand, the EPR spectra of the phenoxyl radical complex, [Cu(MeO-Salcn)]SbF_6, were silent in the temperature range of 4–100 K, which may be due to the large ZFS parameters ($D > 0.3$ cm^{-1}) reported in [44]. The expression "EPR silent" does not specify the detailed electronic structure of the oxidized Cu(II)–phenolate complex, since it could refer to any of the cases, antiferromagnetism, ferromagnetism, and diamagnetism [61]. However, DFT calculation revealed that the two SOMOs consist of the $d_{x2\text{-}y2}$ orbital of copper ion and the ligand *p*-orbital, which are situated in the orthogonal positions. Therefore, magnetic properties of this complex can be assigned to the S = 1 ground state with ZFS parameters D = 0.722 cm^{-1} and E/D = 0.150 based on the *ab initio* calculation [44]. Cu(II)-diiminobenzene radical complex, [Cu(Salphen-OMe)]SbF_6, is predicted to be ferromagnetic

S = 1 ground state due to similar orthogonality between radical and copper d orbitals. Indeed, [Cu(Salphen-OMe)]SbF_6 can be assigned to the ferromagnetic species with S = 1 ground state, based on the pulse EPR experiment [47].

On the other hand, ligand oxidation and metal oxidation can be distinguished in one-electron oxidation of the group 10 metal salen-type complexes. One–electron oxidized group 10 metal salen-type complexes have an unpaired electron with S = 1/2. Metal-centered oxidation species show a large g value and large anisotropy due to coupling with the metal nuclear spin, while the radical species show a g value close to 1.998 for free electron. The EPR spectra of the oxidized Ni(II) salen-type complexes are shown in **Figure 8,** and those of the oxidized group 10 metal Salpn complexes are shown in **Figure 9** and references [40,41,43]. In the general case of low-spin d^7 Ni(III) complexes, the EPR spectrum shows the axial signals with g_{iso} = *ca.* 2.15 [54]. The spectrum of the one-electron oxidized Ni(Salcn) complex shows the signals at g_{av} = 2.05, which is different from the spectrum of the characteristic Ni(III) signals [39-41]. The g values of these oxidized Ni complexes supported formation of the Ni(II)-phenoxyl radical complex as the main species with some contribution from Ni(II) ion, which is in good agreement with the results of the solid state characterizations in [39,40]. Further, [Ni(Salpn)]$^+$ showed the isotropic signal at g = 2.04. Lack of the hyperfine structures based on the nickel nuclear spin suggests that the contribution of the central nickel ion is rather small and that the radical electron is slightly localized on the ligand [43].

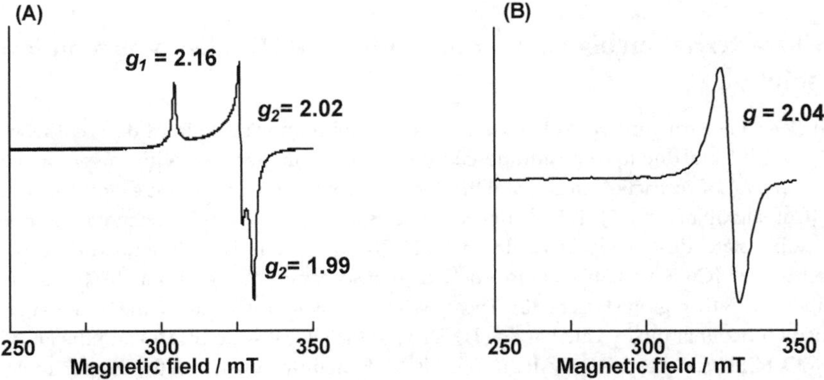

Figure 8. EPR spectra of oxidized Ni complexes in CH_2Cl_2 at 77 K ; (A) [Ni(Salcn)]$^+$ and (B) [Ni(Salpn)]$^+$.

Such a trend is also observed for the Pd and Pt complexes. The one-electron oxidized Pd complexes, [Pd(Salen)]$^+$ and [Pd(Salpn)]$^+$, have a similar characteristic; the band width of the signal is narrow, and the g value is close to the free electron value (g = 2.010 for [Pd(Salen)]$^+$, and g = 2.007 for [Pd(Salpn)]$^+$) in comparison with the other group 10 metal salen-type complexes. The results matches with the results of solid state characterization as relatively localized Pd^{II}(phenoxyl)(phenolate) complexes [40,43]. However, only [Pd(Salcn)]$^+$ shows the EPR signals coupled with Pd ion nuclear spin, indicating that the 6-membered dinitrogen chelate complex has the spin more localized on the ligand [43]. On the other hand, one-electron oxidized Pt complexes, $[Pt(Salen)]^+$ and $[Pt(Salpn)]^+$, exhibited the *g* value similar to that of oxidized Ni complexes, while the band width is significantly large with rhombic component. Such the EPR spectral features are in good agreement with the considerations from the solid state characterization that the oxidized Pt complexes are mainly Pt^{II}-phenoxyl radical species but have a large distribution of the radical electron spin at the Pt ion [40,43].

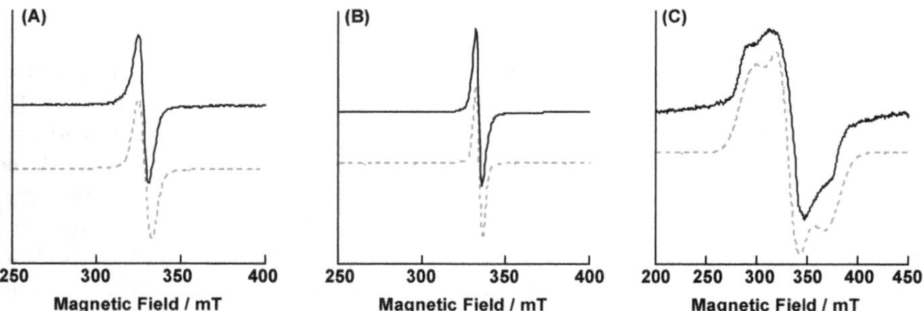

Figure 9. EPR spectra of oxidized group 10 metal Salpn complexes: (A) $[Ni(Salpn)]^+$; (B) $[Pd(Salpn)]^+$ and (C) $[Pt(Salpn)]^+$. Black solid lines are measured spectra and red dashed lines are simulations of measurements.

5. Characterization of one-electron oxidized M(II) salen-type complexes in solution

The electronic structure of oxidized complexes in solution is sometimes different from that in the solid state, due to some additional factors such as interactions with solvent molecules and removal of restriction from the lattice energy. One of the important examples is shown for $[Cu(Salcn)]SbF_6$ in [45]. This complex can be assigned to the Cu(III)-phenolate complex in the solid state described above, but in CH_2Cl_2 solution it is different. The absorption spectrum of $[Cu(Salcn)]SbF_6$ exhibited two intense bands at 18600 and 5700 cm^{-1} in the visible and NIR region (**Figure 10**). These two bands were assigned by the time-dependent density functional theory calculation (TD-DFT) to the HOMO-4 to LUMO and the HOMO to LUMO transition, respectively [62]. The HOMOs mainly consist of the ligand molecular orbitals, while contribution of the copper *d* orbital to LUMO becomes higher. From the result, these two transitions are concluded to be the ligand-to-copper charge transfer (LMCT) bands [45]. However, the LMCT band intensity is temperature dependent, decreasing with increasing temperature (**Figure 10**, inset). Temperature dependent magnetic susceptibility change was also detected by NMR study, and the effective magnetic moment of the CD_2Cl_2 solution of $[Cu(Salcn)]SbF_6$ increased with increasing temperature with 25 % triplet state at 178 K. These temperature dependent changes are reversible, and activation parameters can be estimated to be $\Delta H° = 1.1 \pm 0.1$ kcal mol^{-1}, $\Delta S° = 3.5 \pm 0.1$ cal K^{-1} mol^{-1}, respectively (**Figure 11**). Therefore, the CH_2Cl_2 solution of $[Cu(Salcn)]SbF_6$ showed the valence tautomerism between Cu(III)-phenolate and Cu(II)-phenoxyl radical governed by temperature as reported in [45].

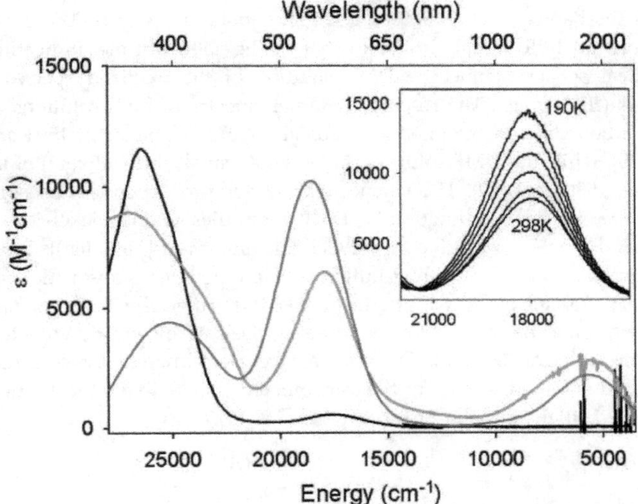

Figure 10. UV-vis-NIR spectra of 0.08 mM solutions of Cu(Salcn) (black) and $[Cu(Salcn)]SbF_6$ (red) in CH_2Cl_2, and the calculated spectrum for singlet $[Cu(Salcn)]^+$ (blue). Inset: Temperature dependence (from 298 to 190 K) of the 18000-cm^{-1} band.

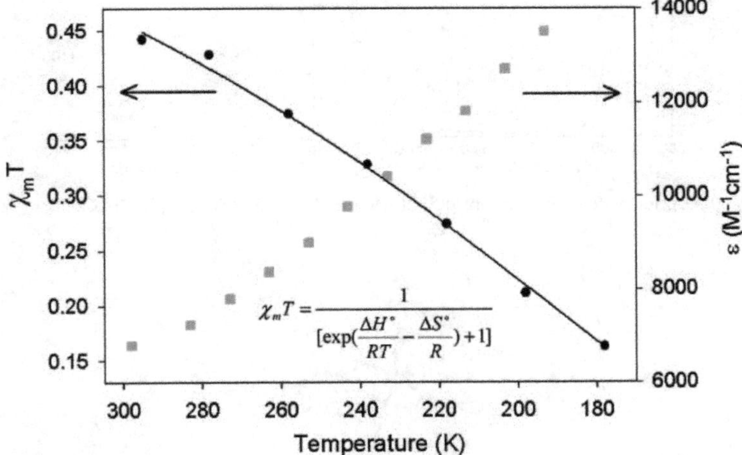

Figure 11. Comparison of the temperature dependent solution susceptibility by ^1H NMR (black circles, CD_2Cl_2) and 18000-cm^{-1} band intensity (red squares, CH_2Cl_2) for $[Cu(Salcn)]SbF_6$. Fitting the susceptibility values (solid line) to the equation indicated affords thermodynamic parameters for the equilibrium.

One-electron oxidized Ni(Salcn) complexes also exhibit a valence state change. The one-electron oxidation of Ni(Salcn) in DMF caused a color change to purple, exhibiting a new absorption band at 476 nm. In CH_2Cl_2, however, a different UV-vis absorption spectrum was detected with the bands at 1100, 900 and 415(shoulder) nm (**Figure 12**) [39]. The CH_2Cl_2 solution showed an EPR signal similar to that of the solid sample, indicating the Ni(II)-phenoxyl radical species, while the DMF solution of the oxidized species showed the characteristic Ni(III) signals. The resonance Raman spectra of both solutions are different, the CH_2Cl_2 solution showed the phenoxyl radical ν_{7a} and ν_{8a} bands at 1504 and 1605 cm^{-1}, respectively [28], while the DMF solution showed only small shifts (3 cm^{-1}) of the phenolate ν_{11a} band at *ca.* 1530 cm^{-1} [39]. The valence state difference dependent on solvents can be considered to be due to coordination of DMF molecules to the nickel ion. Addition of exogenous ligands such as pyridine to the CH_2Cl_2 solution of $[Ni(Salcn)]^+$ causes the color change from green to purple, and the solution showed the UV-vis absorption spectrum and EPR signals characteristic of Ni(III) complexes [63,64]. Further, the X-ray crystal structure of μ-oxo Ni^{III}(salen) dimer has been reported to be synthesized by addition of excess O_2 under basic conditions (**Figure 13**) [55]. Such a valence state change by coordination of an exogenous ligand is also observed for the 6-membered chelate $[Ni(Salpn)]^+$ but not detected for the other group 10 metal salen-type complexes [40,43].

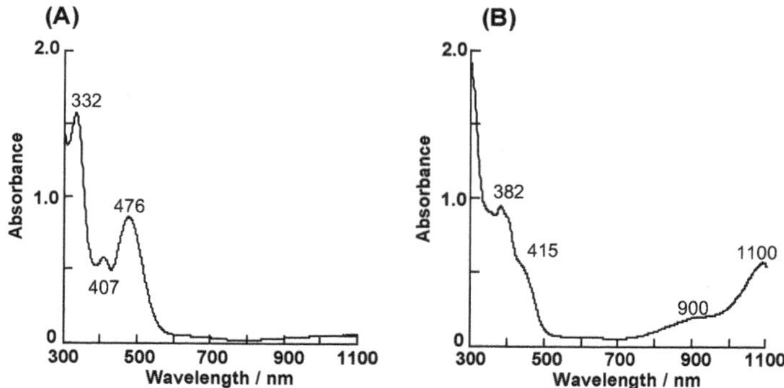

Figure 12. Absorption spectra of solutions of $[Ni(Salcn)]SbF_6$: (A) in DMF; (B) in CH_2Cl_2.

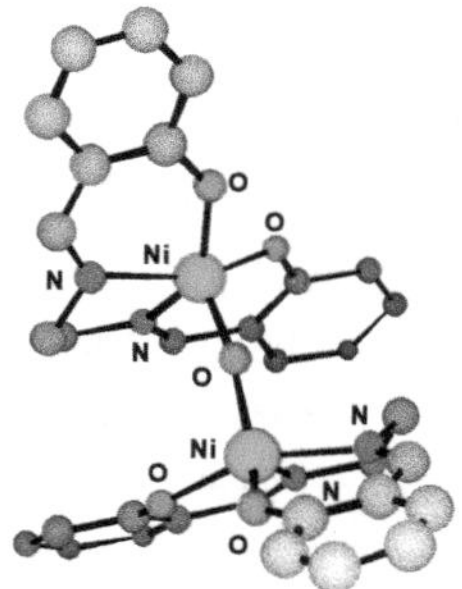

Figure 13. Crystal structure of $[Ni(salen)]_2O$.

One of the important characteristics of the oxidized group 10 metal salen-type complexes, $[M^{II}(Salen)]^+$, $[M^{II}(Salcn)]^+$, and $[M^{II}(Salpn)]^+$, is the appearance of an intense NIR band at ~5000 cm^{-1} [42,43]. The band is assigned to the phenolate to phenoxyl radical (ligand-to-ligand) charge transfer (LLCT) by TD-DFT calculation. The analyses of the low energy NIR LLCT band reveal the degree of the radical delocalization, which can be estimated by Robin-Day classification for understanding the mixed valence system. The classification is categorized in three systems as follows: (1) a fully localized system (class I), (2) a fully delocalized system (class III), and (3) a moderately coupled system (class II) [65]. In the case of the fully localized system, there is no characteristic LLCT band in the NIR region, while the fully delocalized system shows an intense NIR LLCT band. On the other hand, the moderate coupling system exhibits a less intense NIR band, which depends on small perturbations such as solvent polarity. UV-vis-NIR spectra of group 10 metal salen-type complexes are shown in **Figure 14** and reference [43].

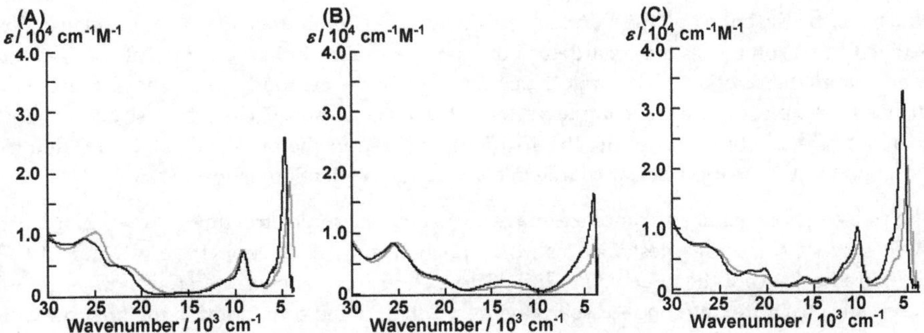

Figure 14. UV-vis-NIR spectra of the one-electron oxidized the group 10 metal salen-type complexes: Salpn complexes, black line; 5-membered chelate complexes, red line. (A) Ni complexes; (B) Pd complexes; (C) Pt complexes.

The NIR band intensity is in the order, Pt > Ni > Pd. The NIR spectrum of the oxidized Pd complex showed a less intense band in comparison to the bands of the oxidized Ni and Pt complexes, and the band was the most solvent dependent. From these results, the oxidized Pd complexes may be slightly closer to the class II moderately coupled system among the oxidized group 10 metal-salen type complexes, that is, the oxidized Pd complex is a more localized system [42]. On the other hand, the NIR band of the Pt complexes was the highest in intensity and less solvent dependent. These results strongly support that the Pt complexes belong to the class III delocalization system, which is also supported by all other experimental results in [42]. As compared with 5-membered and 6-membered chelate dinitrogen backbone complexes, intensity of all of the NIR bands of 6-membered complexes is decreased, indicating that the 6-membered dinitrogen chelate leads to radical localization on the ligand [43]. It is therefore obvious that there is a clear difference between the delocalized and localized systems of the phenoxyl radical species.

6. Conclusion

The oxidation chemistry of the group 10 metal(II) and copper(II) salen-type complexes is discussed in this chapter, to show that the oxidized salen-type complexes have a variety of the oxidation products. The CV of these salen-type complexes exhibited two reversible redox waves of one-electron process and the values of the first redox potential of complexes were in a narrow range. However, one-electron oxidized complexes have different electronic structures, which are dependent on the central metal ion, aromatic ring substituents, and the chelate effect of the dinitrogen backbone.

The electronic structure of the one-electron oxidized group 10 metal salen-type complexes is mainly a metal(II)-radical species, but there is a subtle difference in the detailed electronic structures. The oxidized nickel complexes are the delocalized phenoxyl radical species, but the valence state changes upon addition of an exogenous ligand to form the Ni(III)-phenolate species. Such a valence state change could not be detected for Pd and Pt complexes. The oxidized Pd complexes are relatively localized phenoxyl radical species, which can be described as the Pd(phenoxyl)(phenolate) complexes. On the other hand, the oxidzed Pt(II) complexes are regarded as the fully delocalized phenoxyl radical species and have a large distribution of the radical electron spin on the central Pt ion. Among the group 10 metal complexes, the Pd complexes have the most localized electronic structure. The energy of the *d*-orbitals of group 10 M(II) ions increase in the order Pd < Ni < Pt, due to variation of the effective nuclear charge in combination with relativistic effects.

The one-electron oxidized complexes have different electronic structures showing at least three sorts of complexes described as M(III)-phenolate, M(II)-phenoxyl radical, and M(II)-ligand radical except the phenoxyl radical. These electronic structure differences give rise to the crystal structure differences, especially the position of the counterion in the proximity of the oxidation locus. The bond lengths and angles of oxidized complexes also reveal the electronic structure difference. The magnetic susceptibility, UV-vis-NIR measurements, and other physicochemical data substantiate the electronic structure difference and afford further insights into the novel properties, such as valence tautomerism between Cu(III)-phenolate and Cu(II)-phenoxyl radical in CH_2Cl_2 solution of oxidized Cu(Salcn) complex. It is now obvious that detailed electronic structures of one-electron oxidized complex should be concluded on the basis of the results of various physical measurements.

Many salen-type complexes have been reported also in organic chemistry as the catalysts for organic molecular conversion [17]. However, the detailed reaction mechanism has not been discussed, and especially the electronic structure of the catalyst has been unclear. Information on the detailed electronic structure of the metal ion in complexes may lead to construction of more efficient catalysts and discovery of further interesting phenomena.

Author details

Yuichi Shimazaki
College of Science, Ibaraki University, Bunkyou, Mito, Japan

Acknowledgement

The author is grateful to Prof. Dr. Osamu Yamauchi, Kansai University, for helpful comments and suggestions during preparation of this manuscript.

7. References

[1] Holm R. H, Kennepohl P, Solomon E. I (1996) Structural and Functional Aspects of Metal Site in Biology. Chem. Rev. 96, 2239-2314.

[2] Sono H, Roach M. P, Coulter E. D, Dawson J. H (1996) Structure and Reaction Mechanism in the Heme Dioxygenases. Chem. Rev. 96, 2841-2887.

[3] Meunier B, de Visser S. P, Shaik S (2004) Theoretical Perspective on Structure and Mechanisms of Cytochrome P450 Enzymes. Chem. Rev. 104, 3947-3980.

[4] Rohde J.-U, In J. H, Lim M. H, Brennessel W. W, Bukowski M. R, Stubna A, Münck E, Nam W, Que L. Jr (2003) Crystallographic and Spectroscopic Characterization of a Nonheme Fe(IV)=O Complex. Science, 299, 1037-1039.

[5] Baik M, Newcomb M, Friesner R. A, Lippard S. J (2003) Insights into the P-to-Q conversion in the catalytic cycle of methane monooxygenase from a synthetic model system. *Chem. Rev.*, 2003, *103*, 2385-2420.

[6] Tshuva E. Y, Lippard S. J (2004) Synthetic models for non-heme carboxylate-bridged diiron metalloproteins: strategies and tactics. Chem. Rev. 104, 987-1012.

[7] Costas M, Mehn M. P, Jensen M. P, Que L. Jr (2004) Dioxygen Activation at Mononuclear Nonheme Iron Active Sites: Enzymes, Models, and Intermediates. Chem. Rev. 104, 939-986.

[8] Solomon E. I, Brunold T. C, Davis M. I, Kemsley J. N, Lee S.-K, Lehnert N, Neese F, Skulan A. J, Yang Y.-S, Zhou J (2000) Geometric and electronic structure/function correlations in non-heme iron enzymes. Chem. Rev. 100, 235-350.

[9] Nam, W (2007) Special Issue on Dioxygen Activation by Metalloenzymes and Models Acc. Chem. Res. 40, 465-635.

[10] Petrenko T, George S. D, Aliaga-Alcalde N, Bill E, Mienert B, Xiao Y, Guo Y, Sturhahn W, Cramer S. P, Wieghardt K, Neese F (2007) Characterization of a Genuine Iron(V)–Nitrido Species by Nuclear Resonant Vibrational Spectroscopy Coupled to Density Functional Calculations. J. Am. Chem. Soc. 129, 11053-11060.

[11] Gunay A, Theopold K. H (2010) C–H Bond Activations by Metal Oxo Compounds. Chem. Rev. 110, 1060-1081.

[12] Mirica L. M, Ottenwaelder X, Stack T. D. P (2004) Structure and Spectroscopy of Copper–Dioxygen Complexes. Chem. Rev. 104, 1013-1046.

[13] Lewis E. A, Tolman W. B, (2004) Reactivity of Dioxygen-Copper Systems. Chem. Rev. 104, 1047-1076.

[14] Jörgensen C. K (1969) Oxidation Numbers and Oxidation States. Heidelberg, Springer.

[15] Chirik P. J, Wieghardt K (2010) Radical Ligands Confer Nobility on Base-Metal Catalysts. Science 327, 794-795.

[16] Holland P. L (2008) Electronic Structure and Reactivity of Three-Coordinate Iron Complexes. Acc, Chem. Res. 41, 905-914.

[17] de Bruin B, Hetterscheid D. G. H, Koekkoek A. J. J, Grützmacher H (2007) The Organometallic Chemistry of Rh, Ir, Pd and Pt based Radicals; Higher Valent Species. Prog. Inorg. Chem. 55, 247-253.

[18] Kaim, W (2011) Manifestations of Noninnocent Ligand Behavior. Inorg. Chem. 50, 9752-9765.

[19] Chaudhuri P, Verani C. N, Bill E, Bothe E, Weyhermüller T, Wieghardt K (2001) Electronic Structure of Bis(*o*-iminobenzosemiquinonato)metal Complexes (Cu, Ni, Pd). The Art of Establishing Physical Oxidation States in Transition-Metal Complexes Containing Radical Ligands. J. Am. Chem. Soc. 123, 2213-2223.

[20] Herebian D, Bothe E, Bill E, Weyhermüller T, Wieghardt K (2001) Experimental Evidence for the Noninnocence of *o*-Aminothiophenolates: Coordination Chemistry of *o*-Iminothionebenzosemiquinonate(1-) π-Radicals with Ni(II), Pd(II), Pt(II). J. Am. Chem. Soc. 123, 10012-10023.

[21] Stubbe J, van der Donk W. A, (1998) Protein Radicals in Enzyme Catalysis. Chem. Rev. 98, 705.

[22] Whittaker J. W (1994) Radical copper oxidases. Met. Ions Biol. Syst. 30, 315-360.

[23] Whittaker J. W (2003) Free radical catalysis by galactose oxidase. Chem. Rev. 103, 2347-2363.

[24] Ito N, Phillips S. E. V, Yadav K. D. S, Knowles P. F (1994) Crystal structure of a free radical enzyme, galactose oxidase. J. Mol. Biol. 238, 794-814.

[25] Dyrkacz G. R, Libby R. D, Hamilton G. A (1976) Trivalent copper as a probable intermediate in the reaction catalyzed by galactose oxidase. J. Am. Chem. Soc. 98, 626-628

[26] Hamilton G. A, Adolf P. K, de Jersey J, DuBois G. C, Dyrkacz G. R, Libby, R. D (1978) Trivalent copper, superoxide, and galactose oxidase. J. Am. Chem. Soc. 100, 1899-1912.

[27] Clark K, Penner-Hahn J. E, Whittaker M. M, Whittaker J. W (1990) Oxidation-state assignments for galactose oxidase complexes from x-ray absorption spectroscopy. Evidence for copper(II) in the active enzyme. J. Am. Chem. Soc. 112. 6433-6435.

[28] McGlashen M. L, Eads D. D, Spiro T. G, Whittaker J. W (1995) Resonance Raman Spectroscopy of Galactose Oxidase: A New Interpretation Based on Model Compound Free Radical Spectra. J. Phys. Chem. 99, 4918-4922.

[29] Jazdzewski B. A, Tolman, W. B (2000) Understanding the Copper-Phenoxyl Radical Array in Galactose Oxidase: Contributions From Synthetic Modeling Studies. Coord. Chem. Rev. 200-202, 633-685.

[30] Thomas F. (2007) Ten Years of a Biomimetic Approach to the Copper(II) Radical Site of Galactose Oxidase. Eur. J. Inorg. Chem. 2379-2404.

[31] Shimazaki Y, Yamauchi O. (2011) Recent Advances in Metal-Phenoxyl Radical Chemistry. Indian J. Chem. 50A, 383-394.

[32] Shimazaki. Y, Huth, S, Hirota, S, Yamauchi O (2000) Chemical Approach to the Cu(II)-Phenoxyl Radical Site in Galactose Oxidase: Dependence of the Radical Stability on N-Donor Properties. Bull. Chem. Soc. Jpn. 73, 1187-1195.

[33] Shimazaki, Y, Huth, S, Odani, A, Yamauchi, O (2000) A Structure Model for Galactose Oxidase Active Site and Counteranion-Dependent Phenoxyl Radical Formation by Disproportionation. Angew. Chem. Int. Ed. 39, 1666-1669

[34] Shimazaki, Y, Huth, S, Hirota, S, Yamauchi, O (2002) Studies on Galactose Oxidase Active Site Model Complexes: Effects of Ring Substituents on Cu(II)-Phenoxyl Radical Formation. Inorg. Chim. Acta, 331, 168-177.

[35] Pratt, R. C, Stack, T. D. P (2003) Intramolecular Charge Transfer and Biomimetic Reaction Kinetics in Galactose Oxidase Model Complexes. J. Am. Chem. Soc. 125, 8716-8717.

[36] Pratt, R. C, Stack, T. D. P (2005) Mechanistic Insights from Reactions between Copper(II)–Phenoxyl Complexes and Substrates with Activated C–H Bonds. Inorg. Chem. 44, 2367-2375.

[37] Jacobsen, E. N (2000) Asymmetric Catalysis of Epoxide Ring-Opening Reactions. *Acc. Chem. Res*. 33, 421.

[38] Pfeiffer P, Brietg E, Lubbe E, Tsumaki T (1933) Tricyclische orthokondensierte Nebenvalenzringe. Justus Liebigs Ann. Chem. 503, 84.

[39] Shimazaki, Y, Tani, F, Fukui, K, Naruta, Y, Yamauchi, O (2003) One-electron Oxidized Nickel(II)-Di(salicylidene)diamine Complexes: Temperature Dependent Tautomerism between Ni(III)-Phenolate and Ni(II)-Phenoxyl Radical States. J. Am. Chem. Soc. 125. 10512-10513.

[40] Shimazaki, Y, Yajima, T, Tani, F, Karasawa, S, Fukui, K, Naruta, Y, Yamauchi, O (2007) Syntheses and Electronic Structures of One-Electron-Oxidized Group 10 Metal(II)-(Disalicylidene)diamine Complexes (Metal = Ni, Pd, Pt). J. Am. Chem. Soc. 129, 2559-2568.

[41] Storr T, Wasinger E. C, Pratt R. C, Stack T. D. P (2007) The Geometric and Electronic Structure of a One-Electron-Oxidized Ni(II) bis-(salicylidene)diamine Complex. Angew. Chem. Int. Ed. 46, 5198-5201.

[42] Shimazaki, Y, Stack, T. D. P, Storr, T (2009) Detailed Evaluation of the Geometric and Electronic Structures of One-Electron Oxidized Group 10 (Ni, Pd, and Pt) Metal(II)-(Disalicylidene)diamine Complexes. Inorg. Chem. 48, 8383-8392.

[43] Shimazaki, Y, Arai, N, Dunn, T. J, Yajima, T, Tani, F, Ramogida, C. F, Storr, T. (2011) Influence of the chelate effect on the electronic structure of one-electron oxidized group 10 metal(II)-(disalicylidene)diamine complexes. Dalton Trans. 40, 2469.

[44] Orio M, Jarjayes O, Kanso H, Philouze C, Neese F, Thomas F (2010) X-Ray Structures of Copper(II) and Nickel(II) Radical Salen Complexes: The Preference of Galactose Oxidase for Copper(II). Angew. Chem. Int. Ed. 49. 4989-4992.

[45] Storr T, Verma P, Pratt R. C, Wasinger E. C, Shimazaki Y, Stack T. D. P. (2008) Defining the Electronic and Geometric Structure of One-Electron Oxidized Copper–Bis-phenoxide Complexes. J. Am. Chem. Soc. 130, 15448-15459.

[46] Thomas F, Jarjayes O, Duboc C, Philouze C, Saint-Aman E, Pierre J.-L (2004) Intramolecularly hydrogen-bonded versus copper(II) coordinated mono- and bis-phenoxyl radicals. Dalton Trans. 2662-2669.

[47] Kochem A, Jarjayes O, Baptiste B, Philouze C, Vezin H, Tsukidate K, Tani F, Orio M, Shimazaki Y, Thomas F (2012) One-Electron Oxidized Copper(II) Salophen Complexes: Phenoxyl versus Diiminobenzene Radical Species. Chem. Eur. J. 18, 1068-1072.

[48] Whiteoak C. J, Salassa G, Kleij A. W (2012) Recent Advances with π-conjugated Salen Systems. Chem. Soc. Rev. 41, 622-631.

[49] Dunn T. J, Ramogida C. F, Simmonds C, Paterson A, Wong E. W. Y, Chiang L, Shimazaki Y, Storr T (2011) Non-Innocent Ligand Behavior of a Bimetallic Ni Schiff-Base Complex Containing a Bridging Catecholate. Inorg. Chem. 50, 6746-6755.
[50] Glaser T, Heidemeier M, Fröhlich R, Hildebrandt P, Bothe E, Bill E (2005) Trinuclear Nickel Complexes with Triplesalen Ligands: Simultaneous Occurrence of Mixed Valence and Valence Tautomerism in the Oxidized Species. Inorg. Chem. 44, 5467-5482.
[51] Bordwell F. G, Cheng J. P. (1991) Substituent effects on stabilities of phenoxyl radicals and the acidities of phenoxyl radical cations. J. Am. Chem. Soc. 113, 1736-1743.
[52] Webster R. D (2003) In situ electrochemical-ATR-FTIR spectroscopic studies on solution phase 2,4,6-tri-substituted phenoxyl radicals. Electrochem. Commun. 5, 6-11.
[53] Thomas F (2010) Metal Coordinated Phenoxyl Radicals. in: Hicks, R. G, editor. Stable Radicals: Fundamental and Applied Aspects of Odd-Elelctron Compounds. Wiley, pp. 281-316.
[54] Freire C, Castro B (1998) Spectroscopic characterisation of electrogenerated nickel(III) species. Complexes with N2O2 Schiff-base ligands derived from salicylaldehyde. J. Chem. Soc., Dalton Trans. 1491-1498.
[55] Bag B, Mondal N, Rosair G, Mitra S (2000) The first thermally-stable singly oxo-bridged dinuclear Ni(III) complex. Chem. Commun. 1729-1730.
[56] Barnard C. T. J, Russel M. J. H (1987) Palladium and Platinum. in: Wilkinson G, Gillard R. D, McCleverty J. A editors. Comprehensive Coordination Chemistry; Pergamon, Oxford, Vol. 5.
[57] Connelly N. G, Geiger W. E (1996) Chemical Redox Agents for Organometallic Chemistry. Chem. Rev. 96, 877-910.
[58] Emsley, J (1998) The Elements, 3rd Edition, Oxford, Oxford University Press.
[59] Iwasawa, Y (1995) X-Ray Absorption Fine Structure for Catalysts Surfaces. London, World Scientific Publishing.
[60] Müller J, Weyhermüller T, Bill E, Hildebrandt P, Ould-Moussa L, Glaser T, Wieghardt K (1998) Why Does the Active Form of Galactose Oxidase Possess a Diamagnetic Ground State? Angew. Chem. Int. Ed. 37, 616-619.
[61] Kahn O. (1985) Dinuclear Complexes with Predictable Magnetic Propeties. Angew. Chem., Int. Ed. Engl. 24, 834-850.
[62] Chong D. P (1995) Recent Advances in Density Functional Methods. Singapore, World Scientific.
[63] Rotthaus O, Jarjayes O, Thomas F, Philouze C, Del Valle C. P, Saint-Aman E, Pierre J. –L (2006) Fine Tuning of the Oxidation Locus, and Electron Transfer, in Nickel Complexes of Pro-Radical Ligands. Chem. Eur. J. 12, 2293-2302.
[64] Rotthaus O, Thomas F, Jarjayes O, Philouze C, Saint-Aman E, Pierre J. –L (2006) Valence Tautomerism in Octahedral and Square-Planar Phenoxyl–Nickel(II) Complexes: Are Imino Nitrogen Atoms Good Friends? Chem. Eur. J. 12, 6953-6962.
[65] Robin M. B, Day P (1967) Mixed-valence chemistry: a survey and classification. Adv. Inorg. Chem. Radiochem, 10, 247-422.

Chapter 4

Chromatographic, Polarographic and Ion-Selective Electrodes Methods for Chemical Analysis of Groundwater Samples in Hydrogeological Studies

Ricardo Salgado and Manuela Simões

Additional information is available at the end of the chapter

1. Introduction

The chemical and physical characterization of groundwater and surface water is very important to understand the hydrological and geological dynamic that enriched the water in ions and organic compounds. During the water infiltration and movement into the rocks, the water is subject to numerous interactions between the aqueous and the solid phases through physical, chemical and microbial processes such as dissolution, precipitation, oxidation, reduction, complexation, ad-and desorption, filtration, gas exchange, evaporation, biological metabolism, isotopic redistribution and anthropogenic influences [1]. Groundwater in solution may have a high quantity of inorganic and organic compounds. Its contents are the combined result of the composition of surface water when entering the unsaturated zone of the soil and reactions with minerals in the rock that may modify the water composition. As a result, groundwater contains dissolved solids and gases (CO_2, O_2, H_2S) according to its initial composition, type of the rock, the partial pressure of the gas phase, pH and oxidation potential of the solution. The major ions that can be found in water are chloride, sulphate, bicarbonate, carbonate, sodium, potassium, calcium and magnesium against many others in reduced concentrations (<10 mgL^{-1}) such as iron, manganese, fluoride, nitrate, nitrite, cadmium, lead, chromium, strontium, arsenic and boron. Apart from these natural processes, the water also suffers from contamination by human activities. Solutes, such as heavy metals and organic solvents, are chemically introduced in the water systems mostly in the unsaturated zone. When water is in contact with pore gases contaminants there may be transference between the liquid and the gas states. This is an important way of volatile compounds to migrate from the subsurface. After dissolved in water, these compounds can persist for a long time as a separate liquid phase which has prejudicial effects for the human life and the ecosystems. One of the aims of the water analysis is to obtain better knowledge concerning the water quality, residence times in the aquifer, age, recharge areas, flow paths, and also a potential or prohibitive use due to human pollution problems.

A high number of analytical analysis with several traditional techniques are no longer adequate for this purpose and the development of more green analytical techniques that can measure different ions and organic compounds with the same technique are more suitable. Over the last few decades there has been an increase growth of equipments capable of measuring very low concentrations and also analytical procedures that could concentrate the compounds and increase the signal detected, allowed the hydrogeologist to get more information about the chemical characterization of the groundwater. Equipments, such as chromatography, sensors and microdevices (e.g. microelectrodes), has undergone extraordinary developments. Most of these new analytical instruments have a lower limit to the range in which the results can be quantified and below that range where a compound can be detected but not quantified or as not detected. The quantification limits can be helpful tool for the decision to select the analytical method and equipment for the determination of a specific parameter for the hydrogeological study.

The chromatographic methods applied for the determination ions and organic compounds can be more appropriated in some cases, however the electrochemical techniques such as polarography and voltametry and the ion-selective methods can also be an alternative. The chromatographic methods can be applied to measure ion concentrations such as Cl^-, SO_4^{2-}, NO_3^-, F^-, PO_4^{3-}, Ca^{2+}, Na^+, K^+ and Mg^{2+} (ion chromatography) and to measure organic compounds using liquid chromatography (HPLC) with UV detectors (e.g. photodiode array detector (DAD)) and/or coupled with mass spectrometry such as water soluble pesticides and pharmaceuticals or gas chromatography (GC) to measure volatile organic compounds such as polycyclic musk fragrances. The polarographic and voltametric electrochemical techniques can be an option to measure the concentration of organic compounds (e.g pharmaceuticals) as well as in the same way the ion-selective electrodes for some specific ions in water matrices. Na^+, K^+ , Ca^{2+} , Mg^{2+} can be measured by ion-selective electrodes and in some cases this techniques is more advantageous than the chromatographic.

Pharmaceutical active compounds (PhAC), persistent personal care products (PPCPs) and pesticides are commonly occurring as micropollutants with a potentially significant environmental impact. The growing use of pharmaceuticals is becoming a new environmental problem, as both via human and animal urinary or fecal excretion and pharmaceutical manufacturing discharges, increasing concentrations of pharmaceuticals reach sewage treatment plants (STPs). Due to this extensive use, high concentrations of drugs are found in sewage, depending on their half-lives and metabolism. STP is therefore often ineffective in removing these substances, so that varying concentrations of them can be found in surface and groundwater.

The impact in the environment and public health arises not only from wastewater effluents discharged in aquatic media [2, 3], but also from sludge application in agriculture, since they can desorbs and contaminate the groundwater [4]. PhAC and PPCPs are becoming increasingly recognised as important micropollutants to be monitored in different matrices, groundwater, surface water drinking water and wastewater. The aquifer recharge with treated wastewater can represent an important point source of ions and organic compounds to natural aquatic systems. Due to the fact that analytical approaches normally used to quantify the abundance of these compounds are labour intensive and require various specific procedures, more simplified analytical methods need to be employed for the quantification of pharmaceutically active compounds (PhACs) and polycyclic musks in liquid and solid samples. Many studies have been carried out in different countries and geographical locations [5], the occurrence of PhAC and PPCPs in wastewater and environmental samples is highly dependent on the local diseases, treatment habits and market profiles, thus, the pollution profile and can vary significantly between different countries [5]. PhACs include many families such as antidepressives, anticonvulsants, non-steroidal anti-inflammatory drugs (NSAID), steroidal anti-inflammatory drug (SAID), drugs for asthma and allergic diseases, antihypertensives, β-blockers, lipid regulators, antibiotics, and estrogens [6]. Due to the high diversity of compounds displaying a wide variance of chemical structures, many previous studies have elected to perform a combination of analytical methods targeting specific families of compounds [7, 8]. While this strategy can be advantageous with respect to the analysis of each target group, the time-consuming and labour-intensive nature of the analytical procedures makes a high number of different methodologies undesirable when the goal of the study is to make an overall assessment of PPCPs present in environmental samples. Most of the PhACs can be analysed through High Performance Liquid Chromatography coupled with mass spectrometry HPLC-DAD-MS with the MS working with electro spray ionization in positive (ESI+) or negative (ESI-) mode with the same set of conditions after solid-phase extraction (SPE) using different adsorbent materials according to the neutral or acidic proprieties of the compounds. The musks are non-polar and volatile organic compounds and can be analysed by GC-MS after solid-phase microextraction (SPME) with different extraction fibres [9].

The polarographic and voltametric methods have been widely used for the analysis of organic compounds in samples of natural origin. However, the voltametric methods have not been widely explored for the analysis of many PhAC. The voltametric technique most used for PhAC is the direct current polarography (DCP) and differential pulse polarography (DPP) methods for the analysis of PhAC in water samples [10, 11]. The use of glassy carbon electrode has been suggested for linear sweep and cyclic voltammetric studies for some PhAC such as nifedipine [11]. Adsorptiv cathodic stripping polarographic determination of trace PhAC has been reported with high sensitivity. The detection limit obtained by these methods can be found lower or comparable to other known methods as well as the linearity range obtained. Precision of the method developed implies very low values of relative mean deviation, standard deviation and coefficient of variation. Recovery experiments showed that these methods can be used for quantitative analysis and errors of ±0.2% can be expected. The studies have shown that the polarographic and voltametric methods are simple, reproducible and accurate and can be used to determine many PhAC in the groundwater. Despite the sophisticated instrumentation of analytical tools, complete noninvasive measurements are still not possible in most cases. More often, one or more pretreatment steps are necessary; whose goal is enrichment, clean-up, and signal enhancement during a process of sample preparation [12].

2. Groundwater monitoring plan: Sampling procedure and frequency

The groundwater monitoring plan is determined according to the needs to implement a water quality monitoring program as part of their Source Water Protection Plan. The decision is based, in part, on the high susceptibility of the aquifer and past detections of groundwater contamination and also on the characteristics of the aquifer (confined or unconfined and the soil characteristics (e.g. sand and gravel)). The thin veneer of soil at the ground surface is not a significant confining layer and cannot serve as a barrier to contaminant movement between the ground surface and the aquifer. In addition to the identified aquifer vulnerability, the groundwater contamination by volatile organic compounds (VOC's) can be measured in the source water supply from an unconfirmed spill. Although the detections will show the maximum contaminate levels, and their presence demonstrates the risk of contamination is real. The design of sampling and analysis plan include the top management priorities for developing control strategies in the source water protection plan, such as agricultural chemicals and chemicals associated with auto repair/body shops. Other concerns include underground storage tanks, potential spills along transportation routes, and surface water sources and source of water assessment included in the list of priority contaminants (Directives 2000/60/EC and 2008/105/EC as regards priority substances in the field of water policy). In the groundwater monitoring plan, some groups of the compounds that can be included are:

1. Volatile Organic Compounds (VOCs); a group of potential pollutants that includes many solvents, musk fragrances which are the organic chemical constituents that are most commonly found in groundwater effected by domestic, commercial or industrial operations.
2. Synthetic Organic Compounds (SOCs); particularly those constituents which include the most common herbicides and pesticides and pharmaceutical active compounds found in ground water from human activities such as agriculture and domestic activities.
3. Inorganic Compounds: this will include those chemicals found in ground water most associated with agricultural land use or deicing of roadways. Nitrate and chlorine are the primary contaminants of concern.

Prior to purging and sampling a monitoring well during each monitoring event, the depth to water in the well is measured from a reference point at the top of well casing using an electric ullage tape. Using this measurement to calculate the volume of water in the well, three well volumes will be extracted by bailing. After purging, a sample will be collected for field measurement of the selected indicator parameters; pH, specific conductance, redox potential and water temperature. Results of these measurements will be recorded on the appropriate field sampling form. Next, water samples will be obtained using the dedicated bailer for the required VOC, SOC and inorganic parameters. Upon collection, these samples will be properly labeled and stored in an iced cooler for shipment to the contract laboratory. Finally, one additional sample will be collected from the well and a second field measurement of the indicator parameters will be made. A very large volume of water sample should be collected (e.g 5 L) to analyse organic compounds and it should be collected to plastic (PET) bottles and preserved at 4°C in an isothermal bag during transportation to the laboratory. The purpose of purging – removing water continuously from the well – is to ensure that the water sampled is fresh groundwater and not stagnant casing water, which may differ significantly in quality. The use of bottom loading PVC bailers as the type of sampling equipment to purge their monitoring wells and draw samples is commonly used in the groundwater sampling. This technique is the least expensive to implement and works well with shallow monitoring wells. The purchase of a dedicated bailer for each monitoring well, will eliminating the need for decontamination between sampling events. To insure the removal of casing water and for consistency between sampling events, the technician collecting the water samples will remove three well volumes prior to sample collection. This is a conservative and accepted protocol for groundwater sampling in prolific aquifers. This suggests representative water samples from the monitoring well could be obtained just after one well volume is removed.

The sampling frequency can be defined according to the previous experience of monitoring a specific aquifer. Assuming the results for the first year groundwater monitoring, and near the spring, confirms water quality results in a well within anticipated and acceptable levels, subsequent years monitoring events can be defined as semi-annually. Preferably, sampling of monitoring wells will occur in the spring and fall of each year to better define the water quality of the aquifer at periods of high and low water levels. The collected groundwater samples will be analyzed for the same set of constituents measured during the first year of monitoring.

3. Groundwater sampling and field analysis

In the sampling process of surface or groundwater, it is important to define the purpose of the collecting program, number of samples to be tested, which physical parameters and chemical constituents will be analyzed, as well as where the samples will be collected. It is difficult to obtain samples that accurately reflect the composition of groundwater in the aquifer conditions because the pressure and oxygen concentration change considerably during the sampling process. As a result, the temperature, Eh and pH of the water can change too. Atmospheric oxygen oxidizes components, which are commonly found in anoxic groundwater. Also the degassing of CO_2 will increase the pH causing carbonate precipitation ($CaCO_3$) and concomitant loss of alkalinity.

Most of this type of problems can be overcome by carefully sampling and measuring some parameters in the field such as pH, SC, Eh, and temperature, and a pressurized sample for alkalinity determination. Alkalinity can be rapidly quantified in field by the titration method using a burette or a field alkalinity kit. In some cases, a down-hole determination of temperature, pH, specific conductance, and redox may be needed in conjunction with down-hole sample collection. Hence, any pressure-dependent reaction that will affect the water will result in different values for samples collected in situ and at the well head.

In addition, conservation of samples is typically done to ensure that they retain their physical and chemical characteristics. In the field, it is important to collect samples in clean sample bottles (500 mL high-density polyethylene bottles with polypropylene screw caps), preserve them by cooling, freezing, or acidification immediately after collecting and then storing in a chilled vacuum container in a dark place before transportation to a laboratory for analysis. It is good field practice to clean the sampling device prior to use providing that no residue remains. For that, bottles and devices should be rinsed with a sample of water being sampled to prevent any contamination. After use, acetone and distilled water can be used to rinse thoroughly.

Frequently, nitric acid is added for metals preservation, since it prevents adsorption or precipitation of cations. At the same time, the acidification limits bacterial growth and, as an oxidant, converts ferrous iron to ferric iron and precipitation as FeOOH. Before being analyzed, the samples can usually be preserved for all inorganic compounds during 28 days, at 4°C.

4. Analytical methods for inorganic compounds

The analytical measurement can be either qualitative or quantitative and a very large variety of instruments and techniques can be used for different types of analysis, depending on the cost, information, accuracy, and precision acquired. In most laboratories today, ion chromatography (IC) has replaced older methods of ion analysis because it offers superior sensitivity, accuracy, and dynamic range. Also, it is environment-friendly, extremely fast and versatile. Therefore, this method is particularly advantageous in the analysis of low concentrations as such in high-purity water. IC can also be used for detection and quantification of different ion species in a wide variety of water samples. However, older techniques that do not provide such high results are still in use.

With the purpose of proceeding to the efficient chemical characterization of water and to turn possible the comparison among results, laboratories, hidrogeologists and others have developed sampling protocols. For environmental analysis and determination of inorganic ions in drinking water the EPA (United States Environmental Protection Agency) publishes laboratory analytical methods (most of these methods - 120.1, 130.1, 150.2, 200.7, 206.5, 218.6, 300.1, 365.4, etc. - are published as regulations in the Code of Federal Regulations (CFR) at title *Part 136* and *40 Parts 401-503*), and specifies the type of sample that is needed, the type of sampling container to be used, the method by which the sample container is cleaned and prepared, whether or not the sample is filtered, the type of preservative that is to be added to the sample in the field, and the maximum time that the sample can be held prior to analysis in the laboratory [13].

4.1. Ion chromatographic method

Chromatography is a wide range of physic-chemical separation processes in which the components to be separated are distributed between a stationary and a mobile phase. The name for each of the various types of chromatography depends on the state of aggregation of these two phases. The introduction of high pressure in the separation system and the hardware with software for calculation of the peaks, gas and liquid chromatography have developed into one of the most comprehensive and important methods of modern instrumental analysis. Many ions (anions or cations) in the test sample are separated and quantified quickly and with high precision by an ion chromatographic system containing a guard column, a separator column, with or without suppressor device, and are measured using a conductivity detector. In the technique with chemical suppression, the background conductivity is suppressed both chemically and electronically. In contrast, the direct chromatographic technique employs eluents with salts of organic acids in low concentration on ion exchangers of very low capacity to achieve relatively low background conductivity, which can be suppressed directly by electronic means. This method is applicable to the determination of bromide, chloride, fluoride, nitrate-N, nitrite-N, orthophosphate, sulphate, calcium, potassium, sodium, magnesium and ammonium, in water.

4.2. Potentiometric titration method

The water analysis is not completely done if the carbonate and bicarbonate ions are not determined. Using the alkalinity concept, which is the capability of water to neutralize acids when the presence of calcium and magnesium carbonate ions in it is very high, it is possible to quantify the carbonate and bicarbonate ions. The total alkalinity is the contribution due to all bicarbonates, carbonates, and hydroxides present in the water; and it can be determined by potentiometric titration of an unfiltered sample (100 mL) with a standard solution of strong acid (HCl 0.1 $molL^{-1}$) and phenolphthalein (from pH 8 to pH 10) or methyl orange (from pH 3.1 to pH 4.4) as indicator. The result is expressed as $meqL^{-1}$ of HCO_3^- and CO_3^-, i.e., the volume equivalent of acid added to the water until it changes colour. This method is applicable to all types of water in the range 0.5 - 500 mgL^{-1} alkalinity as $CaCO_3$. The upper range can be extended by dilution of the original sample [14].

4.3. Ion-selective electrode method

Ion-selective electrode methods are regularly used to determine many parameters of water in field and laboratory due to their versatile sensors. They are applicable in many situations for the determination of pH, electrical conductivity (EC), hardness, calcium, sodium, potassium, magnesium and others. The electrodes coupled to a multi-parameter analyzer are designed for the detection and quantify of physical and chemical parameters with calibration for any range of values. For example the potassium ion selective electrode consists of an inert fluorocarbon body with a detachable PVC membrane unit, on the end of which is glued the ion selective membrane. The electrical potential of an ion selective electrode is a function of the activity of certain ions in an aqueous solution. This potential can only be measured against a reference electrode, such as a saturated calomel electrode, placed in the same solution. The electrode should be used in the pH range 4-9.

The problem associated with results obtained with ion-selective methods is the uncertainty caused by instable equilibrium depending on the ionic-strength of the solution. When the electrode is placed in a water sample, the response time may go from 1 to 10 minutes or even much more, and the equilibrium point varies in conformity to the type of electrode and the parameter in measure. In the daily life of a laboratory, the main difficulty is associated with inaccurate results related to the sodium electrode because of interfering ions in the sample. Also, the response time increases considerably because of that. Another issue is the short lifetime of the membrane which, in good performance, does not extend to more than a year if it works every day. In order to produce acceptable measurements, it is highly important that the electrode chosen is in conformity with the sample characteristics.

5. Analytical methods for organic compounds

5.1. Sample preparation for chromatographic and electrochemical methods

The selection and application of the most appropriated analytical technique is related with the proprieties of the organic compound to be identified and quantified in the different samples. Among the proprieties, some of the most important are the acid-base characteristic of the compound, the polarity (polar or non-polar compounds) and the adsorption capacity measured by the octanol-water partition coefficient ($logK_{OW}$) reported for many of the organic compounds. Values of $logK_{OW} < 2.5$ correspond to low adsorption potential, $2.5 > logK_{OW} > 4.5$, to media adsorption potential and $logK_{OW} > 4.5$ to high adsorption potential [15]. The adsorption not only depends of the hydrophobicity but also from the electrostatic forces and pKa of the compound [16]. There is a linear relationship between the $logK_{OW}$ and the pKa of the most of the organic compounds [17]. Organic compound with high adsorption potential are mostly present adsorbed in the solid matrices and need a previous extraction procedure to a liquid phase before the analysis by chromatographic and electrochemical methods. The polar compounds present in water samples can be analysed taking to account the acidic or basic proprieties and most adequate adsorption media for clean-up and pre-concentration technique. The low concentration (ngL^{-1} or pgL^{-1}) mostly frequent of the organic compounds in water samples justifies the previous concentration step by the use of solid-phase extraction (SPE) methods. The non-polar compounds present also in low concentration in water samples need also a previous clean-up and pre-concentration technique by the use of solid-phase micro extraction (SPME) before analysis by chromatographic methods.

5.1.1. *Solid phase extraction (SPE)*

Solid-phase extraction (SPE) is the most used clean-up technique for pre-concentration of water (surface, groundwater) and wastewater samples prior to analysis of the organic compounds. The samples should be previously filtered by 0.45 μm glass fibre membranes (GF 6, <1 μm, diameter 47mm from Wathman, England) and stored at -20°C before analysis by SPE. Different sorbents can be used for clean-up the samples: Oasis HLB (hydrophilic lipophilic balance) cartridges and reverse phase Sep-Pak C18 cartridges, which assures good recovery of compounds in a wide range of polarities. For most of the organic compounds, Oasis HLB and reverse phase (RP C18) as SPE cartridges are the most used in literature due to the polar nature of the compounds and the acidic and neutral characteristics e.g PhAC [18,19]. The selection of the SPE media is directly related with the proprieties of the compound (e.g. acidic or neutral characteristics. RP C18 is most appropriate for the neutral and Oasis HLB general it can be more appropriate for the acidic compounds.

SPE was used for the extraction and clean-up of the liquid wastewater samples. OASIS HLB cartridges (60 mg, 30 μm, Waters, Eschborn, Germany) is used for the acidic organic compounds (e.g acidic PhACs) and RP-C_{18ec} cartridges (500 mg, 50 μm, Waters, Milfort, U.S.) is used for the neutral organic compounds (e.g. neutral PhACs). Each cartridge was previously conditioned with 1 mL methanol followed by 1 mL of Milli-Q water, then dried in a N_2-stream. For the acidic PhACs, 200 mL of filtered water and 10 µl of an internal standard were passed through the OASIS HLB cartridges at pH 2-3. For the neutral, 500 mL of filtered water and 50 µl of internal standard (e.g. meclofenamic acid) is passed through the RP-C_{18ec} cartridges at pH (7-7.5). Samples passed through the SPE cartridges at a flow rate of 20 mL min^{-1} and vacuum pressure of -5 psi, then the cartridges is eluted four times with 1 mL of methanol. The methanol extracts are evaporated to 1 mL by a gentle nitrogen stream. Then, 50 µL of extract are injected into the LC-MS.

For the extraction of the organic compounds (e.g PhAC) adsorbed to the soil, sludges and solid samples, the procedure consists of ultrasonic solvent extraction (USE) using solvents (e.g. methanol/acetone) or pressurized liquid extraction (PLE) using 100% methanol. After this extraction step, non-selective, an additional clean-up can be performed with SPE [19]. The method most commonly used for extraction from solid phase is the ultrasonic solvent extraction (USE) prior to SPE. In this method, the solid sample is centrifuged for 5 min at 10 000 rpm. 2 g of the centrifuged solid sample is for extraction of the organic compounds adsorbed. The concentrated sample is mixed with 4 mL methanol in an ultrasonic bath for 5 min. The slurry is then centrifuged for 1 min at 10 000 rpm. The supernatant is collected in a separate vial and 2 mL of methanol is again added to the solid sample. Centrifugation and supernatant collection is then repeated. To ensure the extraction is complete, 2 mL of acetone is then added to the solid sample and the same procedure (i.e. ultrasonic bath, centrifugation, supernatant collection) is repeated. Then, the 4 extracts (2x2 mL of methanol and 2x2 mL of acetone) are combined and evaporated to a volume of ca. 1 mL. The concentrated extract is diluted in 150 mL of Milli-Q water prior to SPE.

5.1.2. Solid phase micro extraction (SPME)

The most used technique for the determination of the non-polar (e.g polycyclic musk fragrances (PMF)) in water, wastewater, soil and sludge samples of the WWTP is the headspace solid-phase micro extraction (SPME), followed by GC-MS analysis [19]. SPME is also a pre-concentration and clean-up technique previous to analyze by GC-MS. Due to their elevated lipophilicity (logK_{ow} = 5.90-6.35), most of non-polar organic compounds are, therefore, sorbed onto soil or sludge and suspended matter. In literature, analytical methods are reported for analyzing polycyclic musk fragrances (PMF) in soil, sediments and sludge using soxhlet or pressurized liquid extraction (PLE) with dichloromethane, silica gel, alumina columns and gel permeation chromatography (GPC) as clean-up methodology previous to GC-MS analysis [18]. In all the cases, several clean-up steps must be applied to the extracts before chromatographic analysis. The SPME is a solventless technique that simplifies the long and tedious processes of sample preparation and analyte extraction in a single step. The SPME technique is a very sensitive technique that can be applied to adsorb the volatile and non-polar compounds released from the aqueous or solid phase to the headspace completely isolated where a fiber of an adsorbable material or can be immerged in the liquid sample to extract selectively the target compounds [20]. The fibers can be of polyacrilate (PA), polydimethylsiloxane (PDMS), divinylbenbenzene (DVB), PDMS/DVB and carboxen-PDMS (CAR-PDMS) and carbowax-DVB (CW-DVB), carboxen-PDMS-DVB (PDMS-DVB-CAR) and they are selected according to the characteristics of the compound that need to be extracted. The head space technique is more used than the immerged fiber in the liquid phase due to the matrix characteristic of some samples that are inappropriate for the submerged fiber.

The extraction of non-polar organic compounds (e.g. musks) from the water, wastewater, soil, sediment and sludge samples was carried out by solid phase micro extraction (SPME) with fibres previously described. The fibres are pre-conditioned prior to use for 30 min at 250 °C. 2 g of sample is added to a vial with 0.5 g NaCl and 10 μL of an internal standard. The fibre was exposed to the sample headspace in a sealed vial with a Teflon lid for 15 min at 90°C. The fibre was thermally desorbed and analysed by GC-MS.

Sludge is a very complex sample and the extraction of the organic pollutants from the matrix usually implies solvent extraction of the dried soil or sludge samples assisted by accelerated solvent extraction, sonication, microwave heating, solid phase extraction (SPE), simple agitation or solid phase micro extraction (SPME). The determination of non-polar organic compounds in solid samples samples by SPME with different fibers can be influenced by the extraction temperature, fiber coating, agitation, pH and salting out on the efficiency of the extraction. An extraction temperature of 100 °C and sampling the headspace over the stirred sludge sample using PDMS/DVB as fiber coating lead to best effective extraction of the musks in general. The method proposed is very simple and yields high sensitivity, good linearity and repeatability for all the analytes with limits of detection at the ngg^{-1} level. The total analysis time, including extraction and GC analysis, in only 40 min, and no manipulation of the sample is required. The GC-MS with MS in electronic impact (EI) in positive mode analytical technique is the most appropriate for the identification and quantification of the polycyclic musk fragrances (PMF) [9].

5.1.3. Chromatographic analysis of organic compounds

The liquid chromatography coupled with mass spectrometry (LC-MS), liquid chromatography tandem mass spectrometry (LC-MS/MS) and liquid chromatography with diode array detector and coupled with mass spectrometry (LC-DAD-MS) with MS in electrospray (ESI) or atmospheric pressure chemical ionization (APCI) in positive or negative mode are the analytical techniques most adequate for the identification and quantification of organic compounds (e.g PhACs, pesticides, herbicides) in water and wastewater since most of the compounds present neutral and acidic polar characteristics [18]. The GC-MS, LC-DAD-MS and LC-MS/MS techniques are important techniques used for most of organic compounds and their metabolite or reaction products identification and quantification [21, 22]. Due to their selectivity and sensitivity, they are particular important and powerful methods for metabolite or by-product identification of many reactions in the environment (such as biodegradation, photo-oxidation, chemical oxidation and others). Even when a definitive assignment of chemical structures is not possible and, therefore, only tentative degradation pathways can be proposed, GC-MS is so far the most frequently used tool of analysis for identifying transformation products [23]. Winckler used GC-MS for study the ibuprofen metabolites generated by biodegradation processes. Two important advantages of GC-MS methods are the large amount of structural information they yield by the full scan mass spectra obtained under electronic impact (EI) ionization and the possibility of using commercial libraries, making identification of unknowns feasible. However GC-MS has important drawbacks because of its scan capability for analyzing the very polar, less volatile compounds typically generated by these photo-processes [24]. Because of this limitation derivatization techniques should be considered for protection of the polar group by the chemical reaction for a specific period and temperature conditions to get a non-polar derivatized compound that is more compatible with the GC-MS analysis. Many compounds can be used as derivatized reagents to give this protection to the molecule (e.g. MSTFA (*N*-methyl-*N*-trimethylsilyltrifluoroacetamide), BSTFA (*N,O*-bis(trimethylsilyl)-trifluoracetamide), TMS (Trimethylsulfonium hydroxide solution) and MTBSTFA (*N-tert*-Butyldimethylsilyl-*N*-methyl)trifluoracetamide) depending on the chemical structure of the original compound to derivatize. The analysis of degradation products is a highly challenging task. First, the chemical structure of intermediates is unknown, although it can be assumed that primary degradation products are structurally related to the parent pharmaceuticals. Second, standard material for structure elucidation is seldom available. Third, degradation products are present at low concentrations. Therefore, advanced and extended identification methodologies are needed for full structural elucidation of organic compounds (e.g pharmaceutical) degradation products and other new organic compounds that could be found in surface and groundwater as result of chemical and biological transformations. Samples containing the organic compound are typically separated by LC or GC, and either directly injected or pre-concentrated by SPE, lyophilization, evaporation, solvent extraction (e.g. liquid–liquid extraction), or SPME. Many chromatographic techniques can be applied for product isolation prior to nuclear magnetic resonance (NMR). During GC or LC separation, degradation product retention times may provide the first source of identification information. One major point of attention during GC analysis is the thermal stability of pharmaceutical degradation products. High GC-inlet temperatures can decompose thermal labile compounds. They may be used to estimate the polarity and volatility of the degradation products and can be compared, if available, with the standards. For a more accurate identification, degradation product spectral data have to be collected by use of dedicated detection instruments such as UV spectroscopy or mass spectrometry (MS) [22]. When standard compounds are available, LC-UV or GC-MS spectral data of the unknown degradation products are compared with those of standard compounds. GC-MS also allows spectra comparison with extended databases (e.g., from NIST or Wiley). However, in the majority of cases, standards or databases are not available and data on molecular weight, elemental composition, and chemical structure have to be collected by GC-MS, LC-MS, high resolution (HR)-MS, or multidimensional MS (MS^n). Analysis of the parent compound molecule and analogous products as well as isotope labeling strengthens identification. Next to these hyphenated techniques, direct UV photodetection, direct infusion (DI)-MS, and nuclear magnetic resonance (NMR) analysis may provide supplementary data to complement the chromatographic separation.

The MS analysis of the sample is frequently one of the most used techniques for degradation product of target organic compounds identification and also for the biotransformation products of organic compounds studies without analysis of standards, parent molecules, or analogous products. This is possible since the structure of the parent molecule is also known. Another possibility can be as suggested by Doll and Frimmel, in [25] where they made clear distinction between unequivocally identified organic compound (e.g pharmaceutical) degradation products, based on comparison of LC retention time and UV spectrum with standards, and tentatively identified degradation products, based on LC-MS fragmentation analysis with comparison with standards [25]. The proposed techniques can be applied for a wide range of degradation products, i.e., biodegradation and photodegradation products from pharmaceuticals, as well as degradation products from other micro pollutants such as musks. Although standard compounds and GC-MS spectra may be readily obtained for transformation products resulting from AOP treatment or biotransformation metabolites of widely variable PhAC chemical structures and musks studied, these resources are frequently unavailable commercially to confirm the chemical structures for many other organic compounds and some of this standards need to be synthesized in laboratory. Chromatographic separation by GC or LC is an indispensable part of the analytical procedure when multi residue analysis is the focus. The choice between both chromatographic techniques is especially based on the polarity and thermal stability of the target compound.

Another option to confirm the product and metabolite chemical structures is the use of combining the information of LC-DAD-MS and LC-MS/MS. The presence of diclofenac has been reported in natural waters and in wastewater treatment plant effluents as a consequence of its incomplete elimination with conventional wastewater treatment [26, 27]. Direct photolysis can produce photo transformation products that are commonly analyzed by GC-MS, LC-MS and LC-MS/MS techniques combined in order to get confidence in the chemical structures produced during the process. LC-MS and LC-MS/MS allow the separation of semi-polar and polar degradation products without extensive derivatization. Moreover, aqueous samples can be directly injected when concentrations of intermediates are high compared to the instrument detection limit. The retention time of the degradation products can provide information on degradation product polarity. This is a tool for product identification in addition to stronger identification methods.

The mass spectrometry (MS) is considered a principal tool for identifying new products of oxidation of organic compounds, especially since it enables an efficient analysis of trace amounts of analytes in complex organic mixtures [22]. Such complicated environmental or biological samples require separation of components prior to mass spectrometric analysis, which justifies the use of hyphenated techniques such as GC-MS or LC-MS. The single-stage quadrupole (Q) and an ion trap (IT) both hyphenated to a GC, and a quadrupole-time-of-flight (QTOF) MS coupled to an LC can also be used. In the Q mass analyzer, the ions generated in the source undergo electron impact (EI) fragmentation, which results in complex, ambiguous spectral data and hence in non-selectivity that is its main disadvantage [22]. In contrast, the IT mass detector has the unique ability to isolate and to accumulate ions. By iterating ion trapping and scanning, it allows the generation of collision-induced dissociation (CID) spectra of the parent and fragment ions (and their fragment ions), thus increasing the level of confidence in assigning a particular structure [22]. Alternatively, the hybrid QTOF, in which the final resolving mass filter of a triple Q is replaced by a TOF analyzer, not only allows MS^2 operation but also has the necessary accuracy and resolution to give exact-mass measurements [22,28]. Together with MS methods, both chromatographic techniques complement each other to account for a wide range of polarity, acidic-basic characteristics and different functional groups formed during UV degradation or other oxidation techniques. LC-MS methods can also be used for the identification of metabolites produced by organisms like diclofenac in fish bile with electrospray ionization quadrupole-time-of-light mass analyser (QTOF) [28]. The combined use of both GC-MS and LC-MS analysis for detection of organic compounds such as pharmaceutical degradation products targets two different purposes: (a) increasing the range of detectable degradation products or (b) confirmation of suggested degradation products. A large range of degradation products is detected used GC-MS for detection of non-polar degradation products and LC-MS for semi-polar and polar degradation products during advanced oxidation of diclofenac and dipyrone [29].

The improvement of analytical methods confirms that for the majority of the organic trace contaminants, microbial degradation does not lead to mineralization but rather to the formation of a multitude of transformation products. In order to evaluate whether an organic contaminant was transformed to non-toxic products or even mineralized, it is important to know the transformation pathways. Modern hybrid mass spectrometry systems provide the accurate masses of the new products and deliver information of mass fragments which can be used to identify the chemical structure. However, with the exception of very simple reactions (e.g. hydrolysis of amides and esters) the MS spectra are often not sufficient to obtain and confirm the chemical structures of the transformation products [24]. In general, there are a couple of structural modifications which lead to products with the same accurate masses and similar mass fragments of the parent compound. Without the knowledge of chemical/microbial reactions and/or measurements with alternative methods, the suggested product chemical structure could be incorrect. One possible solution for structural confirmation of the transformation products is nuclear magnetic resonance spectroscopy (NMR). However, a drawback of NMR is the elevated quantity needed of a relatively pure isolated standard, not very easy to achieve for the low concentrations.

6. Electrochemical analysis of organic compounds

The pharmaceutical, pesticides and flame retardant are considered emerging organic compounds, some of them are considered xenobiotic compounds. Analytical measurement procedures of these organic compounds can not only fallow LC-MS and GC-MS techniques but also electrochemical analytical techniques. The scope of organic compounds analysis includes the analytical investigation of bulk materials, the intermediates, and degradation products of substances that can be expected to find in the environment resulting from the different urban and rural sources and promote environmental impact in soil, surface and groundwater and consequently in human and health. The presence of these compounds in groundwater and surface water can enter in the urban water cycle and affect the drinking water systems and agriculture when irrigated with contaminated water with the organic compounds. The growing use of pharmaceuticals is becoming a new environmental problem, as both via human and animal urinary or fecal excretion and pharmaceutical manufacturing discharges, increasing concentrations of pharmaceuticals reach sewage treatment plants (STPs). Due to this extensive use, high concentrations of drugs are found in sewage, depending on their half-lives and metabolism. STPs are often ineffective in removing these substances, so that varying concentrations of them can be found in surface and groundwater. In recent years, increasing attention has been paid to the determination of pharmaceuticals, pesticides and flame retardants in water samples. Until now, many analytical methods reported in the literature have been carried out by gas and high-performance liquid chromatography, usually in combination with mass spectrometry (GC–MS, LC–MS), capillary electrophoresis mass spectrometry and high-performance liquid chromatography-photochemically induced fluorimetry (LCPIF). Unfortunately, all these reliable methods are very expensive, and it would be better to use different analytical methods, which do not require expensive instrumentation and which therefore could be used even in less highly developed areas. It is necessary that analytical methods and results comply with the following requirements: 1) the analytical techniques used provide reliable results with a fast turnaround time; 2) the obtained results provided will remain consistent throughout the development cycle of the substances; and if possible, 3) the techniques are transferable to laboratories doing more repetitive testing.

Electrochemistry has always provided analytical techniques characterized by instrumental simplicity, moderate cost and portability. Electroanalytical techniques can easily be adopted to solve many problems of organic compounds with a high degree of accuracy, precision, sensitivity and selectivity, often in spectacularly reproducible. First examples of the organic compound (e.g. pharmaceutical) analysis using by polarographic methods were described in the 1930s and 1940s. Most of the pharmaceutical active compounds (PhAC) were found to be as an electrochemically active. Modern electrochemical methods are now sensitive, selective, rapid and easy techniques applicable to analysis in the pharmaceutical fields, and indeed in most areas of analytical chemistry. They are probably the most versatile of all trace PhAC analysis. Electroanalytical methods are also widely used in specific studies and monitoring of industrial materials, biological and environment samples. The electroanalytical techniques at varying levels of sensitivity are required to solve analytical problems. This kind of assays require high specificity, low detection and determination limits and capable of determining drugs and their metabolites with nanogram (ng) or picogram (pg) level simultaneously. Voltammetric techniques have been extremely useful in measuring drinking water, wastewater, groundwater, surface water, metabolites and urinary excretion of drugs following low doses, especially when coupled with chromatographic methods. In many cases, modern electroanalytical techniques like square wave voltammetry (SWV) can be available alternative to more frequently used spectrometric or separation methods. The volumetric instrument involves a cell with three electrodes immersed in a solution containing the analyte and also an excess of nonreactive electrolyte called supporting electrolyte. One of the three electrode is the working electrode (e.g. microelectrode (ME) of vitreous carbon (VC), or mercury electrodes such as dropping mercury electrode (DME), static mercury electrode (SME) and hanging mercury drop electrode (HMDE)), whose potential varied linearly with time, the other electrode is the reference electrode (commonly a saturated calomel, or a silver/silver chloride electrode (Ag/AgCl/KCl(sat.))) and the third electrode is the counter electrode, which is often a coil of platinum wire or a pool of mercury that simply serves to conduct electricity from the signal source through the solution to the electrode).

6.1. Electrode preparation

The fundamental process in electrochemical reactions is the transfer of electrons between the electrode surface and molecules in the interfacial region (either in solution or immobilized at the electrode surface). The kinetics of this heterogeneous process can be significantly affected by the microstructure and roughness of the electrode surface, the blocking of active sites on the electrode surface by adsorbed materials, and the nature of the functional groups (e.g., oxides) present on the surface. Therefore, there has been considerable effort devoted to finding methods that remove adsorbed species from the electrode and produce an electrode surface that generates reproducible results. Some of these methods have also resulted in the activation of the electrode surface (as judged by an increase in the rate of electron transfer). These methods include mechanical polishing, heat pretreatment, and electrochemical pretreatment. The most common method for surface preparation is mechanical polishing. The protocol used for polishing depends on the application for which the electrode is being used and the state of the electrode surface. There are a variety of different materials available (e.g., diamond, alumina, silicon carbide), with different particle sizes suspended in solution (BAS supplies 0.05 μm alumina polish and 1, 3, 6, and 15 μm diamond polishes). The pad used for polishing also depends on the material being used for polishing - Texmet pads are used with alumina polish, and nylon pads should be used with diamond polish.

Working electrodes supplied by BAS have first been lapped to produce a flat surface, and have then been extensively polished to a smooth, mirror-like finish at the factory. Therefore, they typically only require repolishing with 0.05 μm or 1 μm diamond polish by the user in between experiments. Materials that have a rougher surface (e.g., electrodes which have been scratched) must first be polished using a larger-particle polish in order to remove the surface defects. After the defects have been removed, the polishing should continue with successively smaller-particle-size polish (e.g., 15 μm, then 6 μm, then 3 μm, and then 1 μm). Once polishing has been completed (this can require from 30 s to several minutes, depending upon the state of the electrode), the electrode surface must be rinsed thoroughly with an appropriate solvent to remove all traces of the polishing material (since its presence can affect the electron transfer kinetics).Alumina polishes should be rinsed with distilled water and diamond polishes with methanol or ethanol. The rinsing solution should be sprayed directly onto the electrode surface. After the surface has been rinsed, electrodes polished with alumina should also be sonicated in distilled water for a few minutes to ensure complete removal of the alumina particles. If more than one type of polish is used, then the electrode surface should be thoroughly rinsed between the different polishes. The effect of any surface pretreatment can be determined by its effect on the rate of electron transfer. This can be judged qualitatively by examining the separation of the peak potentials in a cyclic voltammogram of a molecule whose electron transfer kinetics are known to be sensitive to the state of the surface; a more quantitative determination can be made by calculating the value of ks from this peak potential separation. For example, ks for potassium ferricyanide at glassy carbon surface following a simple polishing protocol was found to lie in the range 0.01 - 0.001 cm s^{-1} (this should be compared with the values measured for ks for a platinum electrode, which are at least one order of magnitude larger) [30]. The strong dependence of the electron transfer kinetics of ferricyanide on the state of the electrode surface means that there can be significant variations in the peak potential separation after each polishing. Polishing alters the microstructure, roughness, and functional groups of the electrode surface in addition to removing adsorbed species [30, 31]. For example, the electrode surface can be contaminated by the agglomerating agents required to keep the alumina particles suspended in solution and by the components of the polishing pad. The presence of these species can have a deleterious effect on the electron transfer kinetics by blocking the active sites for the electron transfer reaction. Polishing is often used in combination with another pretreatment (e.g., heat or electrochemical). For many other systems, the simple polishing described above is adequate (for example, when using non-aqueous electrolytes, since blocking of active sites by adsorbed species is less common in such electrolytes than in aqueous solutions). Another method for preparation of the electrode surface that is becoming more widely used is electrochemical pretreatment (ECP), particularly for electrodes which cannot readily be polished (e.g., carbon fiber cylinder electrodes). ECP consists of applying conditioning potentials to the electrode surface before the experiment. As for polishing, this has the effect of removing adsorbed species and altering the microstructure, roughness, and functional groups of the electrode surface. The precise ECP protocol depends upon the application and varies considerably. The potential waveforms typically are held at, or cycle to, a large positive or negative potential, either using steps or sweeps (constant potential, potential scan, triangular wave and square wave. Although the development of the preconditioning protocols has been largely empirical, the pretreated electrode surface has been characterized in order to elucidate the reasons for the activation of the electrode surface [31]. For glassy carbon electrodes, in addition to the removal of adsorbed species, the preconditioning potential leads to the formation of an oxygen-rich layer on the carbon surface. This layer contains oxides as well as other oxygen-containing functional groups which may catalyze electron transfer reactions (the composition of the functional groups in this layer is sensitive to the pretreatment conditions and depends on the solution pH as well as the potentials used for the pretreatment) [31]. The oxide layer can also adsorb and/or exchange ions from the solution, which leads to improved detection limits. However, electrochemical pretreatment of electrodes increases the background current of the electrode relative to that of a polished electrode, which may be disadvantageous for some applications.

6.2. Pre-treatment and enrichment of surface and groundwater water samples for voltametry analysis

A solid-phase extraction (SPE) can also be used for pre-concentration of surface water samples for electrochemical analysis. 2L volume samples were adjusted to pH 3 and 7 with concentrated hydrochloric acid and sodium hydroxide, respectively and then passed through an SPE cartridge (conditioned with acetone, methanol and water) using a vacuum system [32]. After the cartridges had been left to dry for 30 min, the drugs were eluted with 5mL of methanol. The choice of methanol as extracting agent was suggested by its strong eluent ability and its inactivity on the electrodes used. Extracts were diluted with 5mL of 0.2M $NaClO_4$ before electrochemical analysis. In some cases 1mL of the extract obtained at pH 7 was then diluted with 9mL of 0.1M NaOH before analysis [32].

6.3. Ion-selective electrode method

Potentiometric methods also play a significant role in pharmaceutical, biological, clinical and environmental analysis with the introduction of new ion-selective electrodes [33]. It is still one of the most promising analytical tools capable of determining both inorganic and organic substances in pharmaceuticals. Ion selective electrodes (ISEs) belong to the oldest established chemical sensors. Their response characteristics including selectivity, kinetic process and basic thermodynamic process are comparatively well understood. ISEs are mainly membrane based devices, consisting of permeable selective ion-conducting materials, which separate the sample from the inside of the electrode. On the inside is a filling solution containing the ion of interest at a constant activity. The membrane is usually water insoluble, nonporous and mechanically stable. ISEs membranes always contain a substantial concentration of the primary ion salt, which is typically in the mM range. This practice clearly favors the occurrence of co-transport and thus brings about a less than optimal detection limit. The composition of the membrane is designed to yield a potential that is primarily due to the ion of interest via selective binding processes. In addition, the membrane backside is usually contacted with a relatively concentrated internal filling solution of a primary ion salt. These ions can be released into the sample by a zero-current ion flux according to two principal mechanisms. These are the co-transport of the primary ion with a counter-ion and the counter-transport with another ion of the same charge type. The ion transfer mechanisms not only limit the experimental lower detection level but also bias the assessment of the selectivity behavior of the ISE. Selectivity coefficients are typically determined from the responses to the primary and the interfering ions. ISEs are most frequently employed in different variants. The interest in ISEs increased dramatically in recent years since it was shown that the detection limit of conventional ISEs can be lowered towards the picomolar (pM) concentration level, enabling potentiometric trace-level analysis. Membrane materials, possessing different ion recognition properties, have thus been developed to impart high selectivity. The Nernst equation is normally used to describe the ideal response of these cells. To overcome the disadvantages of the ISEs, (slow electrode response, poor selectivity towards other acidic or basic gases, long recovery time, and practical limitations for miniaturization), new concepts have been proposed such as the replacement of the pH glass electrode by a polymeric membrane selective to pH. Depending on the nature of the membrane material used to impart the desired selectivity, ISEs can be divided into three groups: glass, liquid and solid electrodes. Glass electrodes are responsive to univalent cations. Ion-sensitive glass membranes are used in this kind of the ISE. The most common glass electrodes are known as the pH electrode. These electrodes have been widely used for pH measurements. Other types of glass electrodes are used for cation measurements such as sodium, ammonium, and potassium. Liquid-membrane-type ISEs, based on water – immiscible liquid substances impregnated in a polymeric membrane, are widely used for direct potentiometric measurements.

This type of ISE is particularly important because it permits direct measurements of several polyvalent cations as well as certain anions. The membrane-active component can be an ion exchanger or a neutral macrocyclic compound. Ion exchanger and neutral carrier electrodes are the best known liquid-membrane-type ISEs. Most of the ISEs work has been related to the development of solid membranes that are selective primarily to anions. Solid-state ISEs based on immobilization of ion recognition sites in the conducting polymer membrane represent another area of great interest.

The solid state membrane can be made of single crystals, polycrystalline pellets, or mixed crystals such as fluoride, iodide, chloride, bromide, thiocyanide, etc. More directly related to pharmaceutical analysis is the considerable interest devoted to polymeric membranes selective to organic compounds such as drug-active compounds. Considerable work has been devoted to the development of solid membranes that are selective primarily to anions. The solid-state membrane can be made of single crystals, polycrystalline pellets or mixed crystals. Solid-state ISEs contain three major elements. The polymeric matrix, one or several mediators and the active material are constituted these three major elements. PVC is often selected as the polymeric matrix. However aminated and carbocylated-PVC, hydroxylated-PVC, cellulose acetate, silicone rubber, or polyurethane derivatives can also be used as polymeric matrix. The mediators can also be plasticizers. The active materials such as ion pair, ion exchanger or neutral carrier must exhibit high lipophilicity in order to remain in the organic membrane, even upon repeated exposure to aqueous samples. These three major elements must be fully compatible and dissolved in the same hydrophobic solvent. For high membrane sensitivity of this kind of ISE, the active material must rapidly and reversibly exchange the ion. Covalent binding of ion-recognition sites to conducting polymers should result in very durable solid-state ISEs that are easily miniaturized, e.g., electropolymerization of specifically functionalized monomers. The optimal composition of the membrane will be selected in order to provide the best characteristics such as Nernstian slope, the lowest detection limit, the shortest response time, the widest linearity range, the highest selectivity, the best reproducibility, and the longest lifetime. Other types of solid-state ISEs are coated-wire ISEs and solid-state electrodes without an internal filling solution [33]. Coated-wire electrodes are prepared by coating an appropriate film directly onto a conductor thus eliminating the internal filling solutions. It can be prepared simply by dipping the solid substrate into a solution containing the dissolved polymer, the plasticizer and an ion-pair, and allowing the solvent evaporates. The analytical performance of solid-state ion-selective electrodes where conducting polymers are used as ion-to-electron transducer has been significantly improved. This ion-responsive membrane is commonly based on polyvinyl chloride (PVC) while the conductor can be graphite or metallic-based (Pt, Ag, Cu, Ru, Al, etc.) and can be of any convenient geometric shape (wire, disc, etc.). Coated wire electrodes can be prepared by using other polymers and modified polymers such as poly acrylic acid and modified poly vinyl-benzyl chloride, etc. These electrodes are inexpensive, simple and easy to prepare. The measuring concentration range is varied between 10^{-5} and 10^{-1} M. Despite these advantages, coated-wire ISEs may suffer from reproducibility and long-term stability problems because of the poorly defined contact and mechanism of charge transfer between the membrane coating and conducting transducer. New concepts for preparing coated wire ISEs appear to improve their analytical performance particularly with respect to stability and reproducibility. The ability to eliminate the internal electrolyte solution and decrease the detection limits compared with the traditional ISEs are significant advantages. The detection limit of such coated-wire or solid-state electrodes without an internal filling solutions is on the nanomolar (nM) level. Also, these electrodes demonstrate high stability similar to those of conventional ISEs. Possible interferences in ISEs generally originate from compounds that are structurally related to the pharmaceutically active compounds under investigation such as metabolites, intermediates of synthesis, homologues of other pharmaceutically active compounds, or drug excipients exhibiting similar pharmacological properties. The determination and screening of drugs by the method of direct potentiometry with solid-state ISEs offers a rapid and simple procedure satisfying all requirements of pharmaceutical analyses in groundwater and other organic compounds.

6.4. Voltametric and polarographic techniques

Five main potential excitation signals can be used in voltametry such as polarography linear scan voltametry (LSV), normal pulse voltametry (NPV), differential pulse polarography (DPV), square wave voltametry (SWV) and cyclic voltametry (CV) for the analysis of organic compounds in surface water, groundwater and wastewater. Square wave voltammetric (SWV) technique is among the most sensitive means, for the direct evaluation of concentrations; it can be widely used for the trace analysis, especially on pharmaceutical compounds. This method is the source of a fair amount of confusion. The problem arises from the number of waveforms employed, which are frequently described as simply square wave voltammetry. In this context it will be consider three basic groups: the Kalousek, Barker, and Osteryoung formats. Square wave voltammetric technique originates from the Kalousek commutator and Barker's square wave polarography. Kalousek constructed an instrument with a rotating commutator which switched the potential of the dropping. Kalousek square wave technique is a lower frequency method, which measures the current only on the reverse half cycle of the square wave (SW). The Barker format is the simplest to visualize. The waveform is a direct analog to sinusoidal ac voltammetry with a symmetric square wave of frequency and amplitude riding on either a ramp or slow staircase waveform. Osteryoung format is the most common form of SW techniques [34]. This waveform differs from the other SW techniques in that the base potential increases by amplitude for each full cycle of the square wave. The current is measured at the end of each half cycle. This wave form can be applied to a stationary electrode or static mercury drop electrode. In this case the time interval is arranged to allow the drop to grow to a pre-determined size. The response consists of discrete current-potential points separated by the potential increment ΔE. Hence ΔE determines the apparent scan rate, which is a number of current-potential points within a certain potential range. The currents increase proportionally to the scan rate. Frequently, the response is distorted by electronic noise and a smoothing procedure is necessary for its correct interpretation. In this context, it is better if ΔE is as small as possible.

The advantage of SWV is that a response can be found at a high effective scan rate, thus reducing the scan time. For this reason SWV is employed more often than normal pulse voltammetry (NPV) and differential pulse voltammetry (DPV) techniques. Whereas NPV and DPV function with effective sweep rates between 1 and 10 mVs^{-1}, SWV can reach 1 Vs^{-1}. There are advantages: greater speed in analysis and lower consumption of electroactive compounds in relation to DPV, and reduced problems with blocking of the electrode surface. Also, in comparison to both linear sweep and cyclic voltammetry, it as a much broader dynamic range and lower limit of detection because of its efficient discrimation of capacitance current. Analytical determinations can be made at concentrations as low as 10 nM. SWV is 4 and 3 times higher than the DPV response, for reversible and irreversible systems, respectively. Therefore, typical SWV measurements take only 1-5 s whereas DPV requires much longer analysis times at about 2-4 min. [34]. Frequencies of 1-100 cycles per second permit the use of extremely fast potential scan rates. This speed, coupled with computer control and signal averaging, allows for experiments to be performed repetitively and increases the signal-to-noise ratio.

The other advantage of SWV, the difference of current is larger than either forward or reverse currents, so the height of the peak is usually quite easy to read, thus increasing the accuracy. The forward current i_2, reverse current i_1, or difference current ($i = i_2 - i_1$) can be used as the response in this technique. The net current has only very small charging current contributions, and in typical experiments the total faradaic charge is much less than equivalent to a monolayer of material. That is, the system is charged very little by the perturbation. The position and shape of the net current response are remarkably insensitive to size and shape of electrode. A further advantage of the current difference output is that, when the signal lies in the oxygen reduction plateau, the response due to the reduction of oxygen is subtracted out. The sensitivity increases from the fact that he net current is larger than either the forward or reverse components. Also, the sensitivity of SWV is higher than that of NPV and DPV. Square wave voltammetry is a powerful electrochemical technique that can be applied in both electrokinetic and quantitative determination of redox couples strongly immobilized on the electrode surface [31, 34]. In general, computer-based data acquisition may revolutionize the whole area of data collection in electrochemistry since more complex waveforms and current gathering techniques may be employed. This technique requires the power and flexibility of the mini-computer for its development and modern microprocessors for its commercial implementation. Microprocessors can be used to compensate for the practical problem of solution resistance and recently menu-selectable software has been incorporated in a stand-alone instrument which allows background subtraction and signal differentiation. The inherent speed of SWV can greatly increase sample throughput in batch and flow analytical operations. The method can be quite rapid and lends itself to the monitoring of rapid processes such as liquid chromatography. Simplex optimization to maximize peak current by varying the waveform parameters has been examined and SWV has also been used in thin lays. Because of the sensitivity and rapidity SWV is useful for drug analysis in their dosage forms and biological and water samples. The low detection and determination limits permit the analysis of trace amount of drug compound. SWV method was applied to numerous drug active compounds. In addition, SWV detection can also be used to resolve co-elution or co-migrating species for LC and CE methods. Electroanalytical applications of drugs using SWV technique can be consider into direct and stripping measurements. Some pharmaceutical compounds that are analyzed directly, i.e. without accumulation of reactant or product of the electrode reaction. The stripping methods are based either on the accumulation of amalgams and metal deposits, or on the adsorptive accumulation of pharmaceutical compounds and metal complexes.

The hydrodynamic of the voltametry can be obtained by stirring the cell solution with a ordinary magnetic stirrer, by rotating the microelectrode in order to help the analyte flow and migration from the solution to the working electrode under the influence of an electric field, convection resulting from stirring or vibration and diffusion due to concentration differences between the film of liquid at the electrode surface and the bulk solution. All the efforts need to be made to minimize the effect of migration by introduction of an inactive supporting electrolyte. When the concentration of supporting electrolyte exceeds that of the analyte by 50- to 100- fold, the fraction of the total current carried by the analyte approaches zero and result in the rate of migration of the analyte toward the electrode of opposite charge becomes essentially independent of the applied potential.

The applied potential between the microelectrode and the reference electrode can be obtained by the application of the Nernst equation for an electrode reaction of $A + ne^- \leftrightarrow P$:

$$E_{appl} = E_A^o - \frac{0.0592}{n} log \frac{C_P^o}{C_A^o} - E_{ref} \quad (1)$$

Where E_{appl} is the potential applied between the microelectrode and the reference electrode, C_P^o and C_A^o are the molar concentrations of P (product in the bulk solution changed by electrolysis that tends to zero with the reaction in the microelectrode) and A (analyte in the bulk solution unchanged by electrolysis), respectively, in a thin layer of solution at the electrode surface only. It is also assumed that the electrolysis, over short periods of time, does not alter the bulk solution concentration appreciably because the electrode surface is very small.

Some application examples of the electrochemical analysis of pesticide-acaricide (amitraz) [35] and PhAC (nifedipine) [11] are now presented. Electrochemical analysis is used for the determination of amitraz [35]. Amitraz is a formamide acaricide used predominantly in the control of ectoparasites in livestock and honeybees. Amitraz hydrolysis is rapid and occurs under acidic conditions, exposure to sunlight and biodegradation by microorganisms. The main hydrolysis product of amitraz, 2,4 dimethylaniline, is recalcitrant in the environment and toxic to humans. An electrochemical method for the determination of total amitraz residues and its final breakdown product, 2,4 dimethylaniline. Cyclic voltammetry at a glassy carbon electrode showed the irreversible oxidation of amitraz and 2,4 dimethylaniline in the presence of spent cattle dip. A limit of detection in the range of 8.5×10^{-8} M for amitraz and 2×10^{-8} M for 2,4 dimethylaniline is obtained using differential pulse voltammetry. Feasibility studies in which the effect of supporting electrolyte type and pH had on electroanalysis of amitraz and its degradants, showed that pH affects current response as well as the potential at which amitraz and its degradants are oxidised. Britton-Robinson buffer is found to be the most suitable supporting electrolyte for detection of amitraz and its degradants in terms of sensitivity and reproducibility [35]. Studies performed using environmental samples showed that the sensitivity and reproducibility of amitraz and 2,4 dimethylaniline analyses in spent cattle dip is comparable to analyses of amitraz and 2,4 dimethylaniline performed in Britton Robinson buffer. In addition, the feasibility of measuring amitraz and 2,4 dimethylaniline in environmental samples is assessed and compared to amitraz and 2,4 dimethylaniline analyses in Britton-Robinson buffer. Amitraz and 2,4 dimethylaniline is readily detectable in milk and honey. The biological degradation of amitraz and subsequent formation of 2,4 dimethylaniline is readily monitored in spent cattle dip. The breakdown of amitraz to 2,4 dimethylaniline can be monitored using cyclic voltammetry.

Another application of the electrochemical methods has been developed for the trace determination of nifedipine.in a simple and rapid differential pulse polarographic [11].A well-defined single peak with Ep value of -0.51 V is obtained in 0.1M acetate buffer (pH 5.0). The linearity is valid up to 5×10^{-5} M (r =0.9995) with minimum detection limit of 3.5×10^{-8} M. Precision of the method developed is implied from the values of relative mean deviation, standard deviation and coefficient of variation, which are 2.05%, 1.1 and 3.2% respectively. Marketed formulations of nifedipine have been analyzed by calibration and standard addition methods. The studies have shown that the method is simple, reproducible and accurate and can be used in the analysis of this PhAC in water. The use of glassy carbon electrode has been suggested for linear sweep and cyclic voltammetric studies. Adsorptive cathodic stripping polarographic determination of trace nifedipine has been reported with high sensitivity. Anodic electrochemical behavior based on the oxidation of dihydropyridine ring to form pyridine derivative compound has been reported as good signal. However, the detection limit obtained by this method is found to be lower as compared to the known method. Also, linear range is found to be much wider than the reported values.

In recent years, increasing attention has been paid to the determination of pharmaceuticals, pesticides and other organic compounds in water samples. Until now, many analytical methods reported in the literature have been carried out by gas and high-performance liquid chromatography, usually in combination with mass spectrometry (GC–MS, LC–MS). Unfortunately, all these reliable methods are very expensive, and it would be better to use different analytical methods, which do not require expensive instrumentation and which therefore could be used even in less highly developed areas. The electrochemical methods applied to monitoring of pharmaceuticals, pesticides and other organic compounds can be another on determination in surface waters and groundwater as a result of incomplete removal of some of these persistent compounds in sewage and in the wastewater treatment plants.

6.5. Parameters affecting the voltametry

6.5.1. Effect of the ionic strength

Voltammetric measurements in solutions of very low ionic strength, including those without deliberately added supporting electrolyte, became possible by using microelectrodes, electrodes with at least one essential dimension in the range of micrometers or less [36]. Voltammetric measurements without supporting electrolyte departed significantly from the traditional way such measurements were performed. 'Traditional' voltammetry required an excess of electro-inactive ions to make the solution sufficiently conductive, to make a compact double layer, and to suppress migration of electro-active species. Electrode processes at microelectrodes are usually associated with very low currents in the range of nano- (nA) or picoamperes (pA). Consequently, it might seem straightforward that even in solutions of very high resistance, the ohmic IR drop is very low. However, there is one specific phenomenon which contributes to lowering of the ohmic drop. This is an increase in ionic strength in the depletion layer while the electrode reaction of an uncharged substrate advances. The ionic strength increases as a result of the formation of the charged products accompanied by drawing of appropriate amounts of counter ions from the bulk solution. This allows voltammetric measurements in solutions of low conductivity, e.g. solutions without added supporting electrolyte or simply solutions of low support ratio, which is the ratio of the concentration of supporting electrolyte to the concentration of the reactant. It should be mentioned here that it is possible to have both a relatively high conductivity and a low support ratio; this is the case with a very high concentration of electro-active species. For neutral reactants, the use of regular-size electrodes, with sizes in the range of mm, is practically not possible in solutions of low ionic strength. To correct the effect of the ionic strength during the voltammetric analysis, different buffered medium are used with acetate, phosphate, citrate, borate at different pH from 3 to 9 depending on the acidic or basic proprieties of the organic compound being analyzed [34, 37].

6.5.2. Dissolved oxygen effect

Dissolved oxygen is easily reduced at the microelectrode. An aqueous solution saturated with air exhibits two waves attributed to the first reduction of O_2 to H_2O_2 and a second reduction of H_2O_2 to H_2O corresponding to two waves of identical height. The presence of oxygen often interferes with the accurate determination of the other species. The oxygen removal is ordinary the first step in amperometric procedures. The deaeration is obtained by injection of an inert gas (*sparging*) for several minutes. Gaseous nitrogen (N_2) or another inert gas should also be introduced in the cell solution in order to reduce the oxygen effect response in the working electrode and minimize the effect of the oxygen in the voltammogram [37].

6.5.3. pH effect

The half-wave potentials for organic compounds are markedly pH dependent. pH variation may result in a change in reaction product. The electrode process consumes or produces hydrogen ions that will change the pH of the solution at the electrode surface unless the solution is buffered. These changes affect the reduction potential of the reaction and cause drown-out, poorly defined waves. When the electrode is affected by pH changes, nonlinearity in the diffusion current/concentration relationship will encountered. In organic polarography, good buffering is vital for the generation of reproducible half-wave potentials and diffusion currents. The polarograms of the PhAC (e.g nifedipine) were recorded in different buffer systems. Single symmetrical peak was obtained in Britton-Robinson buffer from pH 3.0 to 10.0, acetate buffer from pH 4.0 to 6.0, borate buffer from pH 7.5 to 11.0, Mcllavaine buffer from pH 3.5 to 7.0 and Clark-Lubs buffer from pH 5.0 to 10.0. Acetate buffer at pH 5.0 was selected for all studies as a single sharp peak with high degree of reproducibility was obtained under these conditions. The effect of maxima suppressor was studied using Triton-X- 100, gelatin and bromophenol blue. In the absence of maxima suppressor, the peak was highly unsymmetrical. Addition of gelatin (0.1%, 0.5 ml), bromophenol blue (0.1%, 0.5 ml) or Triton-X-100 (0.1%, 0.5 ml) improves the symmetry of the peak. With addition of 0.5 ml of 0.1% Triton-X-100, a narrow, symmetrical peak was obtained. With every 0.5 ml addition of Triton-X-100 there was about 8% decrease in the diffusion current. Hence, 0.5 ml of 0.1% Triton-X-100 was selected as the optimum concentration for further studies [37].

6.5.4. Functional groups effect

The reactive functional groups of the organic compounds can be expected to produce one or more polarographic waves. The carbonyl groups including aldehydes, ketones and quinones produce polarographic waves. In general, aldehydes are reduced at lower potencials than ketones. The conjugation of carbonyl double bond also results also in lower half-wave potentials. Certain carboxylic acids are reduced polarographically, although simple aliphatic and aromatic monocarboxylic acid is not. Dicarboxylic acid such as fumaric, maleic or phtalic acid in which the carboxylic group is conjugated with one another give characteristic polarograms. Most peroxides and epoxides yield polarograms. Nitro, nitroso, amino oxide azo groups are generally reduced at the dropping electrode. Most organic halogens groups produce a polarographic wave which results from replacement of the halogen group with an atom of hydrogen. Carbon/carbon double bond is reduced when it is conjugated with another double bond, an aromatic ring or an unsaturated group. Hydroquinone and mercaptans produce anodic waves. Most of these functional groups can be detected by the use of dropping mercury electrode (DME) [37, 38].

6.5.5. Solvents effects

The solubility frequently dictates the use of solvents or pure water for organic polarography. Aqueous mixtures containing different amounts of such miscible solvents such as glycols, dioxane, acetonitrile, alcohols, acetic acid can be employed. The supporting electrolytes such as lithium or tetra alkyl ammonium salts can also be used.

6.5.6. Electro deposition step effect

Only a fraction of the analyte is deposited during the electrodeposition step. The quantitative results depend not only upon control of electrode potential but also upon factors as electrode size, length of deposition and stirring rate for both the sample and standard solutions employed for calibration. Microelectrodes for stripping methods are formed from a variety of materials including mercury, gold, silver, platinum and carbon in various forms. The most popular electrode is the hanging mercury drop electrode (HMDE) which consists of a single drop of mercury in contact with platinum wire [37].

7. Chromatografic and electrochemical method validation

7.1. Chromatographic method validation

7.1.1. Calibration curves, limits of detection (LOD) and quantification (LOQ)

Standards are obtained using the stock solutions diluted in methanol. Six-point calibration curve are normally used for each compound, ranging from 5 to 200 ng L^{-1}, depending on the expected concentration range in the surface and groundwater samples. The regression coefficient of the resulting calibration curves should be >0.95 for all compounds. Ten blank samples are analysed by LC-MS (with methanol) and GC-MS (with *n*-hexane) to determine the lowest signal/noise ratio of each analyte. Limits of detection (LOD) for the analytes are calculated by the formula 3xSD/m, where SD is the standard deviation of the lowest signal/noise ratio of the analyte and m is the slope of the calibration curve. Limits of quantification (LOQ) are calculated as 10xSD/m.

Low LOD and LOQ are obtained for the non-polar organic compounds such as musks for Ternes and Salgado in [9, 19], when analysed by GC-MS as well as the recoveries were very high as presented in Table 1 for Smyth and Salgado in [9, 39].

The LOD and LOQ and recoveries obtained water samples of acidic and neutral PhAC, antibiotics and estrogens are presented in Table 2 for the compounds detected in this study. The results showed that the analytical procedure for PhAC enables the detection of a substantial number of pharmaceuticals with LOD and LOQ comparable with many studies and laboratories using the LC-MS(ESI+) for the detection of this organic compounds.

Compound	LOD µM	LOQ µM	Recoveries %	Reference
Cashmeran	4.85E-09 1.02E-7	1.94E-08	83±3	[9] [39]
Celestolide	8.2E-09 6.56E-8	2.46E-08	85±4	[9] [9]
Galaxolide	3.88E-09 4.26E-8	1.55E-08	94±2 82	[9] [19] [39]
Phantolide	4.10E-09 6.97E-8	1.64E-08	97±2	[9] [39]
Tonalide	3.88E-09 3.26E-8	7.75E-09	82±3 78	[9] [19] [c]
Traseolide	7.75E-09 5.04E-9	2.33E-08	85±4	[9] [39]

[9] Salgado et al., 2010; [19] Ternes et al. (2005); recoveries obtained with groundwater; [39] Smyth et al. (2008)

Table 1. LOD and LOQ obtained by GC-MS(EI+) for Musk (non-polar organic compounds)

The estrogens are the main exception, where tandem MS should be used in order to detect these compounds to a concentration that is relevant to assess their potential environmental impact [8]. However, some other compounds, namely carbamazepine, showed lower LOD and LOQ with SPE (RP-C_{18}) followed by LC-DAD-MS(ESI+) when compared to results obtained with SPE (RP-C_{18}) and GC-MS [7]. The LOQ for carbamazepine was similar to that found through LC-MS/MS [8]. In Sacher, three separate analytical methods were used for antibiotics, antibiotics were analysed together with the other acidic compounds [7]. While Sacher [7] found a lower LOD and LOQ for amoxicillin, their recovery was lower than Salgado [9] (36% vs. 83%).

7.1.2. Determination of recoveries in chromatography

For the determination of the percentage of recovery of organic compounds in water samples, the samples are spiked with analytes dissolved in a stock solution (each at 1 mg mL^{-1} methanol). The water samples are spiked with 100 µg L^{-1} of analyte and internal standard (e.g. meclofenamic acid). After spiking, the samples are stirred for homogenisation and to enable a sufficient contact of analytes and standards with the matrix. Relative recoveries are determined in relation to a MilliQ water standard solution, also spiked with 100 µg L^{-1} of analyte and internal standard. Recoveries of the analytes in the individual clean-up steps are determined by SPE for acidic and neutral organic compounds in water matrices as well as MilliQ water, and analysed by LC-DAD-MS. The relative recoveries are calculated from the analyte areas in the influent matrix, subtracting the analyte area quantified in the original unspiked matrix, divided by the area of the MilliQ standard sample, Table 1 and 2.

Compound	LOD µM	LOQ µM	Recoveries %	Ref	Compound	LOD µM	LOQ µM	Recoveries %	Ref
Acidic PhACs					**Neutral PhACs**				
Captopril	2.3E-08	6.91E-08	66±10	[9]	Atenolol	1.13E-08 9.02E-9	3.76E-08 3.08E-8 1.88E-7	>94±5 67 98	[9] [7] [8]
Clofibric acid	6.17E-08 2.19E-8	2.06E-07 7.45E-7 2.07E-6	>98±1 103 82	[9] [7] [8]	Caffeine	1.39E-07	4.69E-07	>82±1	[9]
Diclofenac	2.2E-08 2.74E-8	7.55E-08 9.15E-7 1.58E-6	>79±5 70 89	[9] [7] [8]	Carbamazepine	8.47E-09 4.07E-8	2.97E-08 1.36E-7 4.24E-8	>75±1 74 92	[9] [7] [8]
Enalapril	2.13E-08	7.45E-08	>88±3	[9]	Clorazepate	4.16E-08	1.39E-07	-	[9]
Flurbiprofen	7.38E-08	2.38E-07	>65±1	[9]	Dimethylaminophenazone	1.26E-07 1.86E-8	4.11E-07 6.06E-8 4.33E-7	>83±3 66 93	[9] [7] [8]
Furosemide	5.75E-08	1.91E-07	66±10	[9]	Domperidone	7.04E-09	2.11E-08	-	[9]
Ibuprofen	6.8E-08 1.70E-8	2.23E-07 5.87E-7 2.44E-6	>89±9 110 81	[9] [7] [8]	Etofenamate	5.42E-08	1.82E-07	-	[9]
Indomethacin	1.58E-08 1.22E-8	5.18E-08 4.08E-7 1.13E-6	>89±9 114 90	[9] [7] [8]	Fluoxetine	4.93E-08	1.65E-07	-	[9]
Ketoprofen	8.27E-08 1.89E-8	2.72E-07 6.34E-7 1.98E-6	>96±5 104 94	[9] [7] [8]	Fluticasone	5.62E-08	1.91E-07	>60±6	[9]
Naproxen	7.83E-08 1.66E-8	2.57E-07 5.70E-7 2.18E-6	>86±2 105 91	[9] [7] [8]	Hydroxazine	4.02E-08	1.34E-07	>87±1	[9]
Antibiotics					Indapamide	1.64E-08	4.93E-08	73±1	[9]
Amoxicillin	3.56E-08 1.26E-8	1.18E-07 4.14E-7	>83±4 36	[9] [7]	Nimesulide	4.55E-08	1.49E-07	>86±1	[9]
Ampicillin	8.6E-09	3.15E-08	>67±1	[9]	Paroxetine	7.2E-08	2.37E-07	>82±6	[9]
Estrogens					Ramipril	2.16E-08	7.44E-08	>86±12	[9]
17-α-ethynylestradiol	7.09E-08	2.33E-07	- 76	[9] [8]	Salbutamol	4.6E-08 1.09E-8	1.51E-07 3.81E-8 2.10E-6	>94±1 66 61	[9] [7] [8]
Estrone	6.67E-08	2.22E-07	104±12 82	[a] [8]	Tramadol	6.67E-08	2.23E-07	>86±2	[9]
β-estradiol	1.47E-08	4.41E-08	- 76	[a] [8]					

[9] Salgado et al., 2010; [7] Sacher et al. (2001); recoveries obtained with surface water; [8] Ternes (2001); recoveries obtained with WWTP effluent.

Table 2. LOD and LOQ obtained by LC-MS(ESI+) for PhAC polar (organic compounds)

7.2. Electrochemical method validation: Voltametry

7.2.1. Calibration curves, limits of detection (LOD) and quantification (LOQ)

A calibration curve is obtained by taking different known concentrations of the organic compound in a supporting electrolyte under specific experimental conditions and recording the polarograms and plotting di_p vs. organic compound concentration curve. The wave height/peak height of the polarogram was found to be proportional to the organic compound concentration.

Compound	Electrode type	LOD/LOQ µM	Method	Ref	Compound	Electrode type	LOD/LOQ µM	Method	Ref
Acidic PhACs					**Neutral PhACs**				
Captopril	SMDE HMDE	2.9E-3 2.3E-3	-	[34] [34]	Atenolol	NGITO GCE	0.13 0.16E+3	DPV DPV	[33] [33]
Diclofenac	ion-selective	4.0E-3	Potencio-metry	[33]	Caffeine	DE	-	CV	[33]
Ketoprofen	SMDE	3.9E-7	-	[34]	Nifedipine	GCE	1.1-4.3E-5	CV	[33]
Naproxen	Pt DE	1.04 1.3E-4	CV, LSV DPV	[33] [33]	Nimesulide	CGE	5.0E-8	DPV	[33]
Antibiotics					Propranolol	CPE	2.0E-7	DVP	[33]
Tetracycline	Gold	0.96	CV	[33]	Paroxetine	SMDE	6.2E-5	-	[34]
Azythromycin	CPE GCE	6.2E-7 9.24E-7	SWV DPV	[33]; [34] [33]	Fluoxetine	GCE	1.0	-	[34]
Estrogens					Hydroxyzine	GCE	3.0E-6	-	[34]
17-α-ethynylestradiol	CPE HDME	3.0E-8 5.9E-7	DPV -	[33] [34]	Indapamide	CPE	5.0E-6	DPV/SWV	[33]
Estrone	-	-	-	-	Ramipril	HMDE	1.2	-	[34]
β-estradiol	GCE Gold	4.0E-5 6.6E-8	DPV -	[33] [34]	Salbutamol	GCE	2.5E-7	Amp. Det	[34]
Pesticides					Sertraline	HMDE	1.98E-4	-	[34]
Atrazine	HMDE	0.37	-	[34]	Tramadol	GCE	2.2	-	[34]
Ametryn	Gold	1.08E-7	SWV	[40]					

Abbreviations: CPE: carbon paste electrode; GCE: glassy carbon electrode; HMDE: hanging mercury drop electrode; NGITO: nano-gold particles modified indium tin oxide; SMDE: static mercury drop electrode; DE: Dropping electrode; Amp Det: Amperometric detection. [33] Uslu et al., 2007; [34] Dogan-Topal et al., 2010;[40] Tavares et al., 2005

Table 3. LOD, LOQ and electrochemical method applied to the determination of different organic compounds

The calibration curve of the PhAC (e.g. nifedipine) is obtained by the diffusion current increased linearly with increase in the drop time of 0.4 to 2.0 s. At drop times above 2 s the increase in diffusion current is not linear. With increase in pulse amplitude from 20 to 100 mV, the diffusion current showed a linear calibration plot is obtained from 5.0×10^{-7} M to 5.0×10^{-5} M of nifedipine which gives an equation of regression line with coefficient of correlation 0.9997 indicating high degree of current-concentration linearity in this range. The Ep value is obtained is in the range of -0.47 to -0.60 V, which is attributed to the reduction of nitro group [11]. The lowest determinable limit of nifedipine is found to be 3.5×10^{-8} M. The electrochemical behaviour of acidic and neutral PhACs is studied by cyclic voltammetry and pulse voltammetric techniques on mercury, carbon nanotube paste, carbon paste and gold electrodes. The best results, in terms of sensitivity, linearity range and detection limits, are obtained by differential pulse voltammetry (DPV) for ofloxacin (LOD 5.2 μM), differential pulse polarography (DPP) for clofibric acid (LOD 4.7μM) and normal pulse voltammetry (NPV) for diclofenac (LOD 0.8 μM) and propranolol (LOD 0.5μM). This is obtained when an enrichment step of approximately two orders of magnitude is performed by a solid-phase extraction procedure (SPE) in order to concentrate the samples and tested on spiked river water samples [33, 34].

Pulse techniques such as DPV, DPP and NPV gave the best results with all drugs in terms of sensitivity, linearity range and detection limits. These techniques is then applied to the determination of the same drugs in spiked river water samples, after a preliminary enrichment step of two orders of magnitude based on solid-phase extraction [32].

The main advantages of the method based on SPE/pulse voltammetry are that it can be applied directly to analysis of surface waters without any separation or derivatization of samples, and it is simple, rapid and inexpensive. Unfortunately, the method cannot be applied to analysis of real environmental samples, because its sensitivity does not allow determination of the drugs at their actual concentrations in surface waters (10^{-7} g/L), even after the enrichment step. Nevertheless, as the concentrations of drugs most frequently found in surface waters have been increasing at a dramatic rate – from 10^{-9} to 10^{-7} g/L in the last 20 years – it will be possible, in principle, to use this SPE/pulse voltammetric method as a good alternative to high-cost and time-consuming chromatographic methods.

7.2.2. Recovery percentages

To determine the percentage recovery, a fixed quantity of organic compound sample solution is taken and it is added three different (5, 10 and 15 μg) levels of working standard of the organic compound in MilliQ water and water sample. At each level the polarograms is recorded seven times and the amount of the organic compound is computed using the formula:

$$Percentage\ recovery = \frac{N(\Sigma XY)-(\Sigma X)(\Sigma Y))}{N(\Sigma X^2)-(\Sigma X)^2} \quad (2)$$

Where N is the number of observations, X is the amount of the organic compound added and Y is the amount of the organic compound obtained. The same procedure is adopted for the water samples of organic compound at two different initial concentrations [11].

7.3. Electrochemical method validation: Selective electrodes and ion chromatography

7.3.1. Errors in quantitative analysis: Accuracy and precision

There is a correct value for any measurement, which, nevertheless, remains largely unknown. But, for the purpose of comparison, measurements made by an established method or by an accredited institution are accepted as the true value. All results are subjected to some degree of uncertainty. The difference between the result from the experimental measurement and the true result is called error. The error can be absolute or relative. It is absolute when the numerical difference from the true value is known and it is relative if the error is expressed as a percentage of the measured value.

Errors can occur during water analysis. Depending on the basis of their origin, errors can be of various types: gross, systematic (determinate) and random (indeterminate). Gross error occurs when a measurement is invalidated by a major event, such as the failure of equipment. On the other hand, systemic errors occur when the same experiment is repeated several times and the individual measurements cluster around a mean value. The deviation from the true value is a measure of systematic error and it is often estimated as the deviation of the mean from the true value. They have a Gaussian distribution and equal probability of being above or below the mean. It can be corrected or fixed if the true value of a measurement is known minimizing the difference between the mean of results and the true value or when comparing the same quantity using different analytical methods. Finally, the random errors are associated with the inaccuracies of the worker manipulating an experimental procedure.

Some of the sources of systematic and random errors include improper sampling techniques and handling of sample, mistakes by operators, and inadequate knowledge of a particular experimental procedure, incorrect use, calibration and faulty instruments, erroneous preparation of solutions, and use of contaminated glassware and reagents. Other sources of random error are long experimental techniques, especially when there are fluctuations in the conditions at the various stages of an experiment. Also, the low ionic strength in groundwater when measured with electrode may result in errors since the liquid junction potential across the porous ceramic plug of the calomel electrode varies with the composition to be measured [14].

All measurements have some degree of uncertainty and the error cannot be eliminated completely, although its magnitude and nature can be estimated or reduced with improved techniques. Uncertainty can be estimated with different degrees of precision by experimental measurements and statistical analyses. It is also important to quantify the closeness of an experimental result to the true, generally designated of accuracy. Precision of results is always affected by random error; it can be expressed as the standard deviation of results from their mean obtained from a set of replicates. The mean is the arithmetic average of a set of replicate results.

Almost all analytical process, including sample preparation and analysis, require calibration against standard solutions of known concentrations covering the concentration range expected in the sample. Calibration provides a basis for comparison with a test solution to enable the concentration or other parameter of the test solution to be determined with a high degree of certainty. In water analysis it is common to calibrate equipments using graphical methods and linear regression of internal or external calibration process. The detection limit and the precision of measurements affect the calibration results mainly when the measure is done using the selective electrode. The oscillation during the measurement and the uncertainty of the results is high in this methodology.

7.3.2. Accuracy of chemical analysis by electroneutratily

The concept of accuracy, precision and reliability can be applied to assess the laboratory results. The ability to report what is in the sample is the laboratory accuracy and the ability to reproduce results is the precision. The hydrogeologist can check on the overall procedure of laboratory submitting duplicate samples from the same source or with a known concentration for analysis or requesting the samples to be analyzed in different laboratories. The inter laboratories comparison of results may lead to the detection the problems. To note is that, at low concentrations, there may be huge variations in results of duplicate analysis of water given that the sensitivity of the method is insufficient. Sometimes the incompatibility of different elements found together in a sample can be a warning that something went wrong [41].

The electroneutrality (EN) characteristic of the water is often used as a quick and dirty check on the completeness of the analysis and on the accuracy of the laboratory. The electroneutrality is expressed by equation (3) so that the sum of the charges of all cations must equal the sum of the charges of all the anions present in water solution (principle of electrical neutrality). If the EN is less than 5%, the analysis was correctly done, in terms of ionic balance. When the EN is higher than 5%, the results are not acceptable; either the analysis is erroneous or one or more significant ions were omitted from the analysis. If the difference is greater than 10%, a big error occurred during the process due to reduced concentrations in very pure water (total dissolved solids < 50 mgL^{-1}). In most laboratories, values up to 2% are inevitable. When the EN value is more than 5%, the sampling and analytical procedures should be examined. However, EN is only applicable for major elements since for minor elements it is highly difficult to estimate. For this reason the sum only considers cations Na^+, K^+, Mg^{2+}, and Ca^{2+}, and anions Cl^-, HCO_3^-, SO_4^{2-} and NO_3^- although sometimes other ions such as NH_4^+, Fe^{2+}, H^+, Al^{3+} contribute significantly for the electroneutrality.

$$Electro\ neutrality\ (EN)\% = \frac{(\sum cations - \sum anions)}{(\frac{1}{2}\sum cations + \sum anions)} x100 \quad (3)$$

At the end of the water analysis, the results are examined to assess the consistency of the methodologies used. Table 4 shows the EN applied to data at low and high mineralized water (from 28 μScm^{-1} to 5630 μScm^{-1}) with proposes to evaluate the error in the results.

Sample	(1) EC μScm^{-1}	(2) HCO_3^- $meqL^{-1}$	(3) SO_4^{--} $meqL^{-1}$	(3) Cl^- $meqL^{-1}$	(3) NO_3^- $meqL^{-1}$	Σ anions	(4) Ca^{++} $meqL^{-1}$	(4) Mg^{++} $meqL^{-1}$	(4) Na^+ $meqL^{-1}$	(4) K^+ $meqL^{-1}$	Σ cations	EN %
1	28.5	0.15	0.07	0.19	0.01	0.42	0.20	0.05	0.18	0.00	0.43	2.4
2	33.4	0.15	0.05	0.20	0.01	0.42	0.16	0.00	0.24	0.03	0.43	2.3
3	39.7	0.17	0.09	0.17	0.01	0.44	0.23	0.03	0.18	0.01	0.45	2.2
4	40.0	0.24	0.05	0.26	0.04	0.60	0.23	0.04	0.32	0.02	0.61	1.7
5	69.2	0.15	0.07	0.32	0.20	0.75	0.28	0.09	0.32	0.04	0.73	2.7
6	112.4	0.50	0.26	2.05	0.39	3.20	0.87	0.73	1.63	0.03	3.26	1.9
7	325 457	3.19 3.90	0.61 0.51	0.85 0.57	0.10 0.01	4.75 4.99	2.62* 3.79	0.16	3.17* 0.94	0.52* 0.46	6.31 5.35	28.2 5.8
8	743	6.27	1.52	1.98	0.13	9.9	3.69*		6.13*	0.06*	9.88	0.2
9	997 1080	8.82 5.82	1.89 2.71	3.03 3.38	0.08 0.02	13.82 11.93	3.99* 7.94	0.99	11.6* 3.29	0.12* 0.08	15.71 12.3	12.8 3.1
10	1452	7.45	4.57	4.44	0.71	17.17	5.10*		18.96*	0.15*	24.21	34.0
11	3080	1.85	24.1	12.83	0.03	38.81	7.45*		70.4*	0.49*	78.34	67.5
12	5630	9.26	17.99	14.09	4.35	45.69	8.85*		76.09*	1.45*	86.39	61.6

* Measured with selective electrodes: Consort models 3315B, 3031B, 3041B and 3011B, (1) Measured with multiparameter analyser, Consort, model C833 and electrode S211B; (2) Determined by potentiometric titration method; (3) Quantified using a ionic chromatographic, Dionex, DX-120, columns IonPac AS14 e AG14, eluent 3.5 mM Na_2CO_3/1.0 mM $NaHCO_3$, with chemical suppression and method 300.0 A by EPA; (4) Quantified using a ionic chromatographic, Metrohm, 761 Compact IC, METROSEP Cation 1-2 (6.1010.000), eluent 4 mM $C_4H_6O_6$/1mM $C_7H_5NO_4$, without chemical suppression and method indicated by Metrohm, application bulletin n° 257/1, for the determination of alkaline metals.

Table 4. Electroneutrality (EN) analysis for different groundwater samples

8. Conclusions

The chromatographic techniques for the analysis of inorganic and organic compounds are more expensive but for most of the cases the limits of detection and quantification are lower when compared with the electrochemical methods such as voltammetry or ion selective electrodes. This justifies the use in most of reported results in literature obtained by chromatographic methods when compared with the electrochemical methods. However, in some specific cases and using the best voltammetric technique, it is possible to achieve comparable results of detection for estrogens (e.g. 17α-ethynylestradiol and β-estradiol), antibiotics (e.g azithromycin), pesticides (e.g. ametryn) and PhAC (e.g. propranolol).

Author details

Ricardo Salgado
ESTS-IPS, Escola Superior de Tecnologia de Setúbal do Instituto Politécnico de Setúbal, Campus do IPS, Estefanilha, Setúbal, Portugal
REQUIMTE/CQFB, Chemistry Department, Faculdade de Ciências e Tecnologia da Universidade Nova de Lisboa, Campus de Caparica, Caparica, Portugal

Manuela Simões*
CICEGE, Sciences of Earth Department, Faculdade de Ciências e Tecnologia da Universidade Nova de Lisboa, Campus de Caparica, Caparica, Portugal

Acknowledgement

Fundação para a Ciência e a Tecnologia (FCT, Portugal) and Instituto Politécnico de Setúbal are thankfully acknowledged for financial support through projects PTDC/AMB/65702/2006 and IPS-3-CP-07-2009, respectively, and PEst-OE/EQB/LA0004/2011 and PEst-C/EQB/LA0006/2011, and also by the grant SFRH/PROTEC/49449/2009. Centro de Investigação em Ciências e Engenharia Geológica (CICEGe) for financial support to field work and water analysis.

9. References

[1] Matthess G, Frimmel F Hirsch P, Schulz H D, Usdowski, H-E (1992) Progress in hydrogeochemistry. Springer-Verlag. 544 p.

[2] Peng X, Yu Y, Tang C, Tan J, Huang Q and Wang Z (2008) Occurrence of steroid estrogens, endocrine-disrupting phenols, and acid pharmaceutical residues in urban riverine water of the Pearl River Delta, South China. *Sci Total Environ* 397: 158-166.

[3] Bartelt-Hunt S L, Snow D D, Damon T, Shockley J and Hoagland K (2009) The occurrence of illicit and therapeutic pharmaceuticals in wastewater effluent and surface waters in Nebraska. *Environ Pollut* 157: 786-791.

[4] Carrara C, Ptacek C J, Robertson W D, Blowes D W, Moncur M C, Sverko E and Backus S (2008) Fate of Pharmaceutical and trace organic compounds in three septic system plumes, Ontario, Canada. *Environ Sci Technol*, 42: 2805-2811.

[5] Zuccato E, Castiglioni S, Fanelli R, Reitano G, Bagnati R, Chiabrando C, Pomati F, Rossetti C and Calameri D (2006) Pharmaceuticals in the environment in Italy: causes, occurrence, effects and control. *Environ Sci Pollut Res* 13(1): 15-21.

[6] Langford K H and Thomas K V (2009) Determination of pharmaceutical compounds in hospital effluents and their contribution to wastewater treatment works, *Environmental International*, 35: 766-770.

* Corresponding Author

[7] Sacher F, Lange F T, Brauch H and Blankenhorn I (2001) Pharmaceuticals in groundwater, Analytical methods and results of a monitoring program in Baden-Wurttemberg, Germany. *J Chromatogr A*, 938: 199-210.

[8] Ternes T (2001) Analytical methods for the determination of pharmaceuticals in aqueous environmental samples. *Trends in Analytical Chemistry*, 20(8): 419-434.

[9] Salgado R, Noronha J P, Oehmen A, Carvalho G, Reis M A M (2010) Analysis of 65 pharmaceuticals and personal care products in 5 wastewater treatment plants in Portugal using a simplified analytical methodology. *Water Sci Technol* 62(12): 2862-2871.

[10] Thakur K, Pitre K S, (2008) Polarographic (DCP & DPP) Determination of ellagic acid in strawberries & pharmaceutical formulations, *J Chinese Chem Soc* 55: 143-146.

[11] Jeyaseelan C, Jugade R, P Joshi, A P (2011) Differential Pulse Polarographic determination of nifedipine in pharmaceutical formulations, *Int J Pharm Sci Drug Res* 3(3): 253-255.

[12] Mitra S (2003) Sample preparation techniques in analytical chemistry. New York. John Wiley & Sons. 488 p.

[13] Fetter C W (1994) Applied Hydrogeology. Third Edition. Prentice Hall, New Jersey. 691 p.

[14] Nollet L M L L (2000) Handbook of water analysis. New York: Marcel Dekker, Inc. 921 p.

[15] Joss A, Keller E, Alder A C, Gobel A, McArdell C S, Ternes T, Siegrist H (2005) Removal, of pharmaceutical and fragrances in biological wastewater treatment, *Water Res*, 39: 3139- 3152.

[16] Metcalfe C D, Miao X S (2003) Distribution of acidic and neutral drugs in surface waters near sewage treatment plants in the lower Great Lakes, Canada, *Environ Toxicol Chem* 22(12): 2881-9.

[17] Caliman F A, Gavrilescu M (2009) Pharmaceuticals, Personal Care Products and Endocrine Disrupting Agents in the Environment - A Review, *Clean - Soil, Air, Water* 37(4-5): 277-303.

[18] Ternes T A, Herrmann N, Bonerz M, Knacker T, Siegrist H, Joss A (2004) A rapid method to measure the solid-water distribution coefficient (Kd) for pharmaceuticals and musk fragrances in sewage sludge, *Wat. Res* 38: 4075-4084.

[19] Ternes T, Bonerz M, Herrmann N, Loffler D, Keller E, Lacida B B and C Alder A C (2005) Determination of pharmaceutical, iodinated contrast media and musk fragrances in sludge by LC/tandem MS and GC/MS. *J Chromatogr A* 1067: 213-223.

[20] Carpinteiro J, Quintana J B, Rodríguez I, Carro A M, Lorenzo R A, Cela R (2004) Applicability of solid-phase microextraction followed by on-fiber silylation for the determination of estrogens in water samples by gas chromatography–tandem mass spectrometry, *J. Chromatogr A*, 1056(1-2): 179-185.

[21] Agüera A, Perez-Estrada L A (2005) Application of time-of-flight mass spectrometry to the analysis of phototransformation products of diclofenac in water under natural sunlight. *J. Mass Spectrom.* 40(7): 908-915.

[22] Kosjek T, Perko S, Heath E, Kralj B, Zigon D (2011) Application of complementary mass spectrometric techniques to the identification of ketoprofen phototransformation products, *J Mass Spectrom* 46: 391-401.

[23] Winkler M, Lawrence J R, Neu T R (2001) Selective degradation of ibuprofen and clofibric acid in two model river biofilm systems. *Water Res* 35(13): 3197-3205.

[24] Ternes T, Wick A., Prasse C (2011) Transformation of emerging micro pollutants in biological and chemical treatment: a challenge for the future?, Proceedings of 8th IWA leading edge conference on water and wastewater technologies, 6th-10th June, Amsterdam,wastewater issue, IWA-009, p.1-4.

[25] Doll T E, Frimmel F H (2003) Fate of pharmaceuticals—photodegradation by simulated solar UV-light, *Chemosphere* 52: 1757-1769.

[26] Xue W, Wua C, Xiao K, Huang X, Zhou H, Tsuno H, Tanaka H (2010) Elimination and fate of selected micro-organic pollutants in a full-scale anaerobic/anoxic/aerobic process combined with membrane bioreactor for municipal wastewater reclamation, *Water Res* 44: 5999-6010.

[27] Jélic A, Gros M, Ginebreda A, Cespedes-Sánchez R, Ventura F, Petrovic M, Barcelo D (2011) Occurrence, partition and removal of pharmaceuticals in sewage water and sludge during wastewater treatment, *Water Res* 45: 1165-1176.

[28] Kallio J M, Lahti M, Oikari A, Kronberg L (2010) Metabolites of the aquatic pollutant diclofenac in fish bile, *Environ. Sci. Technol.* 44: 7213-7219.

[29] Pérez-Estrada L A, Malato S, Gernjak W, Aguera A, Thurman E M, Ferrer I, Fernndez-Alba A R (2005) Photo-Fenton Degradation of Diclofenac: Identification of Main Intermediates and Degradation Pathway, *Environ. Sci. Technol.* 39: 8300-8306.

[30] McCreery R L, Kline K K (1995) Laboratory Techniques in Electroanalytical Chemistry, 2nd Edition. In: P.T. Kissinger and W.R. Heineman editors. Dekker, New York, Chap. 4 and 10.

[31] Alemu H, Hlalele L (2007) Voltammetric determination of chloramphenicol at electrochemically pretreated glassy carbon, *Bull. Chem. Soc. Ethiop.* 21(1): 1-12.

[32] Ambrosi A, Antiochia R, Campanella L, Dragone R, Lavagnini I (2005) Electrochemical determination of pharmaceuticals in spiked water samples, *J. Hazard Mater* 122: 219–225.

[33] Uslu B, Ozkan S A (2007) Solid electrodes in electroanalytical chemistry: present applications and prospects for high throughput of drug compounds, *Combined chemistry & high throughput screening* 10: 495-513.

[34] Dogan-Topal B, Ozkan S A & Uslu B (2010) The Analytical Applications of Square Wave Voltammetry on Pharmaceutical Analysis, *The Open Chemical and Biomedical Methods Journal* 3: 56-73

[35] Brimecombe R D (2006) Voltammetric analysis of pesticides and their degradation: a case study of amitraz and its degradants. *Masters thesis*, Rhodes University.

[36] Ciszkowska M, Stojek Z (1999) Voltammetry in solutions of low ionic strength. Electrochemical and analytical aspects, *J Electroanal Chem* 466: 129–143.

[37] Skoog D A, Holler F J, Nieman T A (1998) Principles of Instrumental analysis, 5th edition, Sauders College publishing, Orlando, Florida, Cap. 25, pp 639-670.

[38] Zuman P (2006) Principles of application of polarography and voltametry in the analysis of drugs, *J. Pharm. Sci.* 31: 97-115.

[39] Smyth S A, Lishman L A, McBean E A, Klywegt S, Yang J J, Svoboda M L, Seto P (2008) Seasonal occurrence and removal of polycyclic and nitro musks from wastewater treatment plants in Ontario, Canada. *J Environ Eng Sci*, 7: 299-317.
[40] Tavares O, Morais S, Paiga P, Delerue-Matos C (2005) Determination of ametryn in soils via microwave-assisted solvent extraction coupled to anodic stripping voltametry with gold ultramicroelectrode, *Anal. Bioanal.Chem.* 382: 477-484.
[41] Appelo, C A J, Postma D (1996) Geochemistry, groundwater and pollution. A. A. Balkema, Rotterdam. 536 p.

Chapter 5

Potentiometric Determination of Ion-Pair Formation Constants of Crown Ether-Complex Ions with Some Pairing Anions in Water Using Commercial Ion-Selective Electrodes

Yoshihiro Kudo

Additional information is available at the end of the chapter

1. Introduction

Ion-pair formation equilibrium-constants or association ones in water have been determined so far using various methods. As representative methods, one can suppose conductometry, spectrophotometry [1], potentiometry [1], voltammetry [2], calorimetry, electrophoresis [3], and so on. The conductometric measurements generally have high accuracies for their determination for metal salts (MX_z at z = 1, 2) and of their metal complex-ions (ML^{z+}) with pairing anions (X^{z-}) in water and in pure organic solvents. According to our knowledge, its experimental operation requires high experimental know-how to handle the measurements. Also, the spectrophotometric measurements require the condition that either species formed in or those consumed in the ion-pair formation are of colored at least. Solvent extraction methods are generally difficult to establish some experimental conditions, such as ionic strength (I) of both phases and solvent compositions, compared with the above two methods. Strictly speaking, its constants are hard to recognize as thermodynamic ones.

We treat here the ion-pair formation of crown compounds (L), such as 15-crown-5 and 18-crown-6 ethers (15C5 and 18C6), with colorless alkali, alkaline-earth metal ions, and so on in water [4-8]. The methods described above are difficult to apply for the determination of the constants. For example, conductometry cannot distinguish among the metal ions M^{z+}, their ML^{z+}, and X^{z-} in water and ML^{z+} is unstable in many cases. Also, many M^{z+} and ML^{z+} employed here cannot be detected spectrophotometrically. Voltammetric methods cannot apply for the determination, because working electrode suitable for M^{z+} detection is difficult to get. While, polarography with DME can be effective for the measurements of such systems [2]. Unfortunately, it must use mercury and its salts which pollute the environment around us.

Thus, in order to overcome these limitations, potentiometry with ISE has been applied for the determination of the ion-pair formation constants ($K_{MLXz}{}^0$) for MLX_z in water at $I \to 0$ mol dm^{-3}, although its applications are limited by kinds of commercial ISE. In the present chapter, its fundamentals and applications for the formation systems of MX_z and MLX_z in water are described. Here, the determination of $K_{MXz}{}^0$, the ion-pair formation constant of MX_z in water at $I \to 0$, is always required for that of $K_{MLXz}{}^0$.

2. Emf measurements

2.1. Electrochemical cells [4-8]

Constitutions of cells employed for emf measurements of test solutions are described as follows.

As a cell with a single liquid junction

Cell (A): Ag|AgCl|0.1 mol dm^{-3} $(C_2H_5)_4NCl$, LiCl, NaCl, KCl,

or 0.05 mol dm^{-3} $MgCl_2$|* test solution|ISE [4,5]

or as that with a double liquid junction

Cell (B): Ag|AgCl|0.1 mol dm^{-3} $(C_2H_5)_4NCl$, NaCl,

or KCl|1 mol dm^{-3} KNO_3|* test solution|ISE [6-8].

Here, the 1 mol dm^{-3} solution of KNO_3 in Cell (B) is a salt bridge, between which and the test solution, E_j estimated from the Henderson equation [2] is in the range of 1 to 3 mV in many cases [9]. Standard types for the reference electrodes of Cells (A) and (B) are Ag|AgCl|0.1 mol dm^{-3} $(C_2H_5)_4NCl$ and Ag|AgCl|0.1 mol dm^{-3} KCl|1 mol dm^{-3} KNO_3, respectively. For Cell (A), the E_j values are corrected by the Henderson equation {see Eq. (1)}, while they are not corrected for Cell (B).

2.1.1. *For ion-selective electrodes*

Commercial ISEs used here are summarized in Table 1 and some comments for the present emf measurements are described.

2.1.2. *Corrections of liquid junction potentials [2]*

For emf measurements of the electrochemical cells, the problem of the liquid junction potentials E_j occurred at the interface marked with an asterisk cannot be avoided. Hence, correction procedures of E_j are described in this section. Here, the salt bridges with KNO_3 are experimentally used and the Henderson equation [2]

ISE	Type & Maker	Comments
Na^+-selective electrode	1512A-10C, Horiba	Glass membrane; this electrode partially responds K^+ and Rb^+.
K^+-selective electrodes	1200K, Toko	Glass membrane (not produced now); this electrode partially responded Li^+.
	8202-10C, Horiba	Liquid membrane; it is unstable for the aqueous solution of hydrophobic picrates.
Ag^+-selective electrode	8011-10C, Horiba	Solid membrane
Cd^{2+}-selective electrode	8007-10C, Horiba	Solid membrane; it responds Br^- partially and I^- well. See below.
Cl^--selective electrode	8002-10C, Horiba	Solid membrane
Br^--selective electrode	8005-10C, Horiba	Solid membrane
I^--selective electrode	8004-10C, Horiba	Solid membrane

a. The above ISEs are used with a Ag | AgCl reference electrodes, Horiba, types 2660A-10T (single junction type) and 2565A-10T (double junction one).

Table 1. Commercial ISEs used here[a]

$$E_{\mathrm{j}} = \phi^{\mathrm{II}} - \phi^{\mathrm{I}} = -\frac{\sum_j |z_j| \frac{u_j}{z_j}(c_j^{\mathrm{II}} - c_j^{\mathrm{I}})}{\sum_j |z_j| u_j (c_j^{\mathrm{II}} - c_j^{\mathrm{I}})} \frac{RT}{F} \ln \frac{\sum_j |z_j| u_j c_j^{\mathrm{II}}}{\sum_j |z_j| u_j c_j^{\mathrm{I}}} \tag{1}$$

is analytically employed for the correction of E_{j}. For example, the molar concentration of j in the phase I, c_j^{I}, consists of $c_{\mathrm{K}}^{\mathrm{I}}$ and $c_{\mathrm{Cl}}^{\mathrm{I}}$, while c_j^{II} does of $c_{\mathrm{Na}}^{\mathrm{II}}$, $c_{\mathrm{NaL}}^{\mathrm{II}}$, and $c_{\mathrm{Pic}}^{\mathrm{II}}$ for the NaPic-L system, where ions involved in a part of the cell are expressed as

$$K^+, Cl^- \text{ (phase I)} \,|^{*}\, Na^+, NaL^+, Pic^- \text{ (phase II)} \tag{2}$$

(see the case 1L). Also, the $u_{\mathrm{Na}}c_{\mathrm{Na}}^{\mathrm{II}} + u_{\mathrm{NaL}}c_{\mathrm{NaL}}^{\mathrm{II}}$ term in Eq. (1) can be assumed to be $u_{\mathrm{Na}}(c_{\mathrm{Na}}^{\mathrm{II}} + c_{\mathrm{NaL}}^{\mathrm{II}})$ in practice (see Table 2 for the u_{Na} value), because the condition of $c_{\mathrm{Na}}^{\mathrm{II}} \gg u_{\mathrm{NaL}}c_{\mathrm{NaL}}^{\mathrm{II}}/u_{\mathrm{Na}}$ holds in many cases with L (low stabilities of ML^{z+} in water).

2.2. Emf measurements

2.2.1. Preparation of calibration curves

A representative procedure for preparing a calibration curve is described below. Using a pipette, 55 cm^3 of the aqueous solution of NaCl or KCl is precisely prepared in 100 cm^3 beaker, kept at 25 ± 0.3 or 0.4 °C, and then slowly stirred with a Teflon bar containing a magnet. To this solution, the ISE corresponding to Na^+ or K^+ and the reference electrode are immersed. A 7 or 10 cm^3 portion of pure water is added by the pipette and, after 5 minutes, emf values at a steady state are read. This operation is repeated, when the amount of the solution is reached to about 100 cm^3. Consequently, 7 or 5 data are obtained in a unit operation. The thus-obtained calibration curves are shown in Fig. 1.

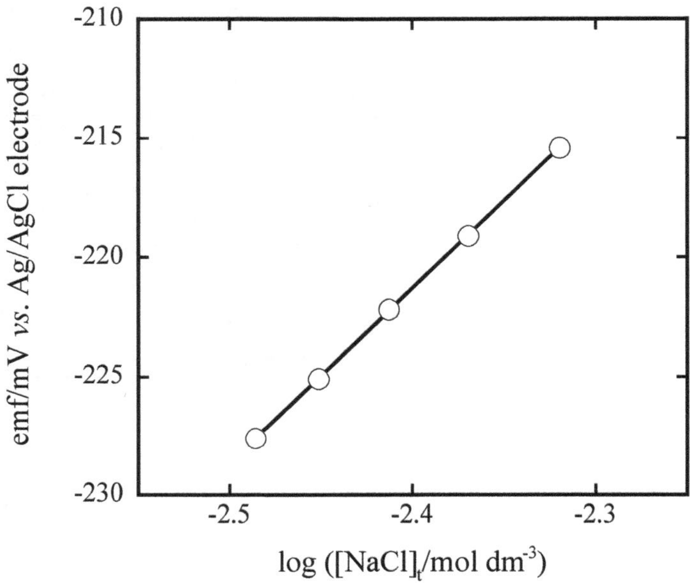

Figure 1. Calibration curve of aqueous NaCl solution at 25 °C. Emf = 55.0log $[NaCl]_t$ – 89.4 at R = 0.999. This slope is close to 59 mV/decade, showing the Nernstian response with z = 1

A high-purity NaCl (99.98 to 99.99%) is dried about 160 °C in an oven [12]. The purities of other standards are checked by $AgNO_3$ titration for which the Ag^+ solution is standardized with the high-purity NaCl.

2.2.2. *Emf measurements of test solutions*

A representative procedure for the emf measurements of the test solution is described here. Using a pipette, 55 cm^3 of aqueous solution of MX_z is precisely prepared in 100 cm^3 beaker, kept at 25 ± 0.3 or 0.4 °C, and then stirred with the Teflon bar. To this solution, the ISE and the reference electrode are immersed. A 7 or 10 cm^3 portion of the aqueous solution of L (or pure water for the K_{MXz} determination) is added by a pipette and after 5 minutes, emf values at a steady state are read. This operation is repeated, when the amount of the solution is reached to about 100 cm^3. Consequently, 7 or 5 data are obtained in a unit operation. Then an example for the plot is shown with that of the corresponding calibration curve of the $CdSO_4$-18C6 system (Fig. 2). These plots indicate that $[Cd^{2+}] \ll [CdSO_4]_t$ in mixtures, compared with the calibration curve depicted at open circles. All the above experiments are performed under the condition of $[MX_z]_t \approx [L]_t$.

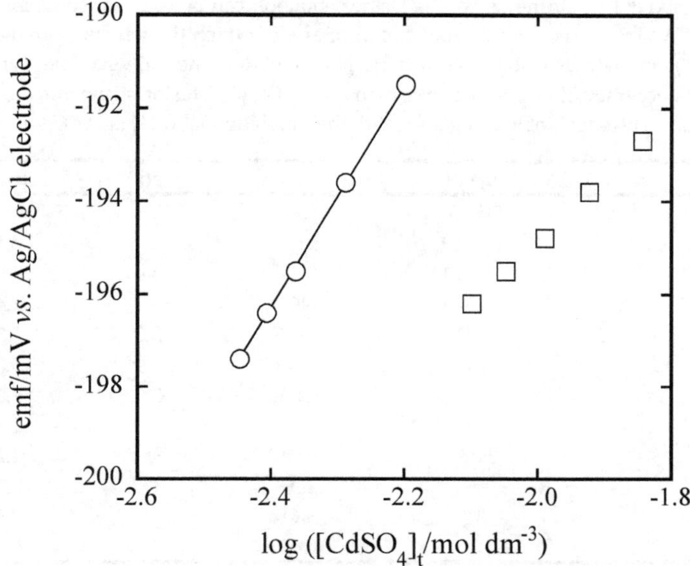

Figure 2. Calibration curve (open circles) of aqueous $CdSO_4$ solution at 25 °C. Emf = 23.8log $[CdSO_4]_t$ − 89.4 at R = 0.999. This slope is close to 30 mV/decade, showing the Nernstian response with z = 2. Open squares show plots of mixtures of $CdSO_4$ with 18C6 in water at 25 °C.

3. Theoretical treatments and data analysis

3.1. Fundamentals

3.1.1. Ionic activity coefficients [10,13]

In order to determine the ion-pair formation constants at $I \to 0$ and 25 °C, the ionic activity coefficients (y_j) of ionic species j used for the activity (a_j) calculations are evaluated from the extended Debye-Hückel equation

$$\log y_j = -\frac{0.5114 z_j^2 \sqrt{I}}{1 + 0.3291 a \sqrt{I}} \tag{3}$$

and the Davies one [13]

$$\log y_j = -0.5114 z_j^2 \left(\frac{\sqrt{I}}{1 + \sqrt{I}} - 0.3I \right), \tag{4}$$

where a denotes the ion size parameter (see Table 2). In general, it is mentioned that the former equation is employed in the range of less than 0.1 mol dm^{-3}, while the latter one is done in that of less than about 1 mol dm^{-3}. Also, the Davies equation can be used for some ions, such as ML^{z+} and $M^{II}X^{+}$, or for the y_j calculations of species j, of which the ion size parameters (e.g., $DDTC^{-}$, tfa^{-}) are not available, because its equation does not involve the parameter a. However, the accuracy of its y_j will be less than that of Eq. (3). The ion size parameters of some ions in water are listed in Table 2, together with their mobility data [11] at 25 °C.

j	a(j) /Å	u_j /10^{-4}	j	a(j) /Å	u_j /10^{-4}
H^+	9	36.25	Cl^-	3	7.912
Li^+	6	4.010	Br^-	3	8.13
Na^+	4	5.193	I^-	3	7.96
K^+	3	7.619	NO_3^-	3	7.404
Ag^+	2.5	6.41$_5$	MnO_4^-	3.5	6.35
Ca^{2+}	6	6.166	ReO_4^-	3.9[c]	5.69$_7$
Cd^{2+}	5	5.6	ClO_4^-	3.5	7.05
$CdCl^+$	4	---[d]	Pic^-	7	3.149
$(C_2H_5)_4N^+$	6	3.38$_4$	BPh_4^-	9[c]	2.2
			SO_4^{2-}	4	8.27
			$S_2O_3^{2-}$	4	8.81
			CrO_4^{2-}	4	8.8

a. Ref. [10].
b. cm^2 s^{-1} V^{-1} unit. Calculated from ionic conductivity data. See Refs. [2] and [11].
c. Calculated from the Brüll formula. See Ref. [10].
d. Not be available.

Table 2. Ion size parameters[a] and mobilities[b] of some ions in water at 25 °C

3.1.2. Model of ion-pair formation equilibria in water [4,6-8,12]

We introduce here three kinds of chemical equilibria for the ion-pair formation of single MX, MX_2, and M_2X and their mixtures with L, except for chemical ones for the mixture of M_2X with L.

Case (1). 1:1 and 2:2 electrolytes

$$M^{z+} + X^{z-} \rightleftharpoons MX \text{ at } z = 1, 2 \tag{5}$$

Case (2). 2:1 electrolytes

$$M^{2+} + X^{-} \rightleftharpoons MX^{+} \tag{6-1}$$

$$MX^{+} + X^{-} \rightleftharpoons MX_2 \tag{6-2}$$

Case (3). 1:2 electrolytes

$$M^{+} + X^{2-} \rightleftharpoons MX^{-} \tag{7-1}$$

$$M^{+} + MX^{-} \rightleftharpoons M_2X \tag{7-2}$$

Case (1L). 1:1 or 2:2 electrolytes with L

$$M^{z+} + X^{z-} \rightleftharpoons MX \tag{8-1}$$

$$M^{z+} + L \rightleftharpoons ML^{z+} \tag{8-2}$$

$$ML^{z+} + X^{z-} \rightleftharpoons MLX \quad \text{at } z = 1, 2. \tag{8-3}$$

Case (2L). 2:1 electrolytes with L

$$M^{2+} + X^{-} \rightleftharpoons MX^{+} \tag{9-1}$$

$$MX^{+} + X^{-} \rightleftharpoons MX_2 \tag{9-2}$$

$$M^{2+} + L \rightleftharpoons ML^{2+} \tag{9-3}$$

$$ML^{2+} + X^{-} \rightleftharpoons MLX^{+} \tag{9-4}$$

$$MLX^{+} + X^{-} \rightleftharpoons MLX_2 \tag{9-5}$$

Ion-pair formation of M^{+} or X^{2-} for Case (3) is omitted, because an example for the ion-pair formation of ML^{+} with X^{2-} was not found for the present experiments. First, one determine the formation constants for Case (1) or (2) and next do those for Case (1L) or (2L), using the equilibrium constants determined by analyzing Case (1) or (2), respectively. Therefore, as you know, the experimental errors of the *K* values obtainable from Cases (1) and (2) become influenced those in the *K* determination of Cases (1L) and (2L), respectively (see Table 5).

3.2. Theoretical treatments and data analysis for Case (1) [4]

Analytical equations are derived from the models represented in the section 3.1.2. Using these equations, analytical procedures are described for the cases of the metal salts MX, M_2X, MX_2, and their mixtures with L, except for those of M_2X with L. Here, M^{z+} and X^{z-} at $z = 1$ and 2 denote a metal ion and a pairing (or counter) anion, respectively.

3.2.1. *Mass and charge balances and the theoretical treatments*

To solve the above equilibria, mass- and charge-balance equations are shown. As an example, Case (1) is described as follows [4].

$$[MX]_t = [M^{+}] + [MX] + b_s \text{ for the species with } M^{+} \tag{10}$$

$$[MX]_t = [X^{-}] + [MX] + b_s \text{ for those with } X^{-} \tag{11}$$

$$\text{and} \quad [M^+] = [X^-] \tag{12}$$

For Case (1), its ion-pair formation constant (K_{MX}) in molar concentration unit is defined as

$$K_{MX} = y_M y_X K_{MX}^0 = \frac{[MX]}{[M^+][X^-]} \quad (y_{MX} = 1) \tag{13}$$

Considering an apparent total concentration to be $[MX]_t - b_s$, one can express [MX] and $[X^-]$ as functions of $[M^+]$. Thus, taking logarithms of the both sides of Eq. (13) and rearranging its equation, the following one is obtained.

$$\log \frac{K_{MX}}{y_\pm^2} = \log\left(K_{MX}^0 + \frac{b_s}{a_M^2}\right) \tag{14}$$

$$\text{with } y_\pm = \sqrt{y_+ y_-} \tag{14-1}$$

$$\text{and } K_{MX} = ([MX]_t - [M^+])/[M^+]^2. \tag{14-2}$$

When the $[M^+]$ value is determined with ISE, then Eq. (14-2) is easily obtained at a given I.

3.2.2. Data analysis [4]

Hence, one can plot $\log (K_{MX}/y_\pm^2)$ against a_M^2 and immediately obtain the K_{MX}^0 and b_s values (mol dm^{-3} unit) from analyzing its plot by a non-linear regression. Figure 3 shows the plot for the NaPic system at 25 °C [4]. Table 3 lists the analytical equations for the other cases, together with equations expressing I which are derived from the charge-balance equations. Also, details for calculation of the parameters listed in Table 3 are summarized in Table 4 [4,6,8,12].

Case	Equation for analysis	Plot for analysis	I /mol dm^{-3}
(1)	$\log \frac{K_{MX}}{y_\pm^2} = \log\left(K_{MX}^0 + \frac{b_s}{a_M^2}\right)$ at z = 1, 2; b_s/mol dm^{-3}: curve-fitting parameter	$\log (K_{MX}/y_\pm^2)$ *versus* a_M^2	$[M^+]$ or $[X^-]$
(2)	$\sum a_j/a_M = 1 + K_1^0 a_X + K_1^0 K_2^0 (a_X)^2$	$\sum a_j/a_M$ *versus* a_X	$[M^{2+}] + [X^-]$
(3)	$\sum a_j/a_X = 1 + K_1^0 a_M + K_1^0 K_2^0 (a_M)^2$	$\sum a_j/a_X$ *versus* a_M	$[M^+] + [X^{2-}]$
(1L)	$\log \frac{K_{MLX}}{y_\pm^2} = \log\left(K_{MLX}^0 + \frac{b_c}{y_{ML}([X^-]-[M^+])a_X}\right)$ at z = 1; b_c/mol dm^{-3}: curve-fitting parameter	$\log (K_{MLX}/y_\pm^2)$ *versus* $y_{ML}([X^-] - [M^+])a_X$	$[X^-]$
(2L)	$F(a_j) \approx 1 + K_{MLX}^0 a_X + K_{MLX}^0 K_{MLX2}^0 (a_X)^2$ with $F(a_j) = \frac{\sum a_j - a_M(1 + K_{MX}^0 a_X)}{a_{ML}}$	$F(a_j)$ *versus* a_X	$[M^{2+}] + [ML^{2+}] + [X^-]$

Table 3. Equations and plots for the equilibrium analyses and I expressions corresponding to them [4,6,8,12]

Case	Molar concentrations of respective species at equilibrium	Remarks
(1)	$[M^{z+}]$ or $[X^{z-}]$: analyzed for $z = 1, 2$	Ref. [4]
(3)	$[M^+]$: analyzed; $[X^{2-}] = \dfrac{[M^+]}{2 + K_{MX}[M^+]}$; $[MX^-] = \dfrac{K_{MX}[M^+]^2}{2 + K_{MX}[M^+]}$; $[M_2X] = [M_2X]_t - \dfrac{[M^+](1 + K_{MX}[M^+])}{2 + K_{MX}[M^+]}$, where $K_{MX} = [MX^-]/[M^+][X^{2-}]$	Ref. [12]
(1L)	$[M^{z+}]$: analyzed; $[X^{z-}] = \dfrac{[M^{z+}](1 + K_{ML}[M^{z+}])}{1 - K_{ML}K_{MX}[M^{z+}]}$; $[ML^{z+}] = [X^{z-}] - [M^{z+}]$; $[MLX] = [MX]_t - [X^{z-}](1 + K_{MX}[M^{z+}])$, where $K_{MX} = [MX]/[M^{z+}][X^{z-}]$ for $z = 1, 2$	Ref. [4]
(2L)	$[X^-]$: analyzed; $[M^{2+}] = \dfrac{[MX_2]_t - \sum[MLX_z]}{1 + K_{MX}[X^-] + K_{ML}[L]}$; $[L] = \dfrac{[L]_t - \sum[MLX_z]}{1 + K_{ML}[M^{2+}]}$; $\sum[MLX_z] = [MX_2]_t - [M^{2+}](1 + K_{MX}[X^-] + K_{ML}[L])$, where $K_{MX} = [MX^+]/[M^{2+}][X^-]$	One can compute $[M^{2+}]$, $[L]$, and $\sum[MLX_z]$ by a successive approximation. See Ref. [8] for its details.

a. Equations correspond to Cases in Table 3. See the text for Case (2)

Table 4. Equations and plots for the equilibrium analyses and I expressions corresponding to them [4,6,8,12]

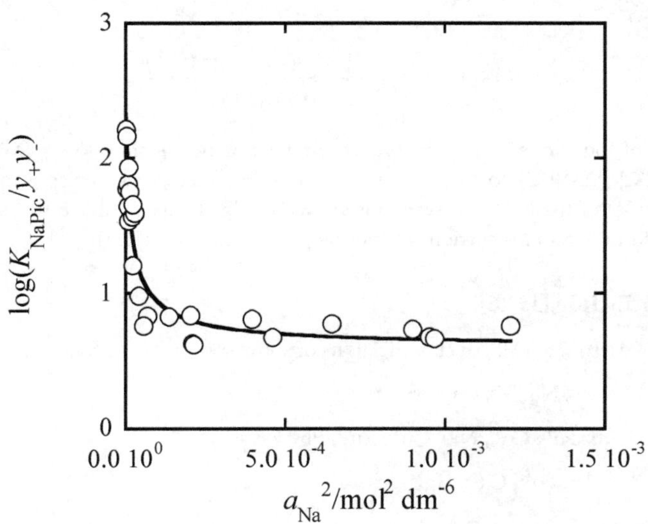

Figure 3. Plot of log ($K_{NaPic}/y_{\pm}^2$) *versus* a_{Na}^2 with $b_s = 5.3 \times 10^{-4}$ mol dm^{-3} and $R = 0.945$ [4].

3.2.3. 2:2 electrolytes [6]

Table 3 lists the equation [6] for the equilibrium analysis and the parameters for its analytical plot corresponding to Case (1) at $z = 2$. Similarly, the equations for the calculation of the parameters and equilibrium constants are summarized in Table 4.

3.3. Theoretical treatments and data analysis for Case (2) [12]

From the mass- and charge-balance equations [12] of Case (2), the following equations are derived.

$$[X^-] = \frac{2[M^{2+}]}{1 - K_{MX}[M^{2+}]} \quad \text{for } M^{2+} \text{ determination by ISE} \tag{15}$$

or

$$[M^{2+}] = \frac{[X^-]}{2 + K_{MX}[X^-]} \quad \text{for } X^- \text{ determination by ISE} \tag{16}$$

The concentrations of other species are

$$[MX^+] = \frac{2K_{MX}[M^{2+}]^2}{1 - K_{MX}[M^{2+}]} \tag{17}$$

and

$$[MX_2] = [MX_2]_t - \frac{[M^{2+}](1 + K_{MX}[M^{2+}]^2)}{1 - K_{MX}[M^{2+}]}. \tag{18}$$

On the basis of the above equations, one can immediately calculate $K_{MX} = [MX^+]/[M^{2+}][X^-]$ and $K_2 = [MX_2]/[MX^+][X^-]$ for a given I. For the other cases, see Tables 3 and 4. As an example, the plot of the Na_2CrO_4 system is shown in Fig. 4. Also, Table 5 lists the $K_{MX}{}^0$ and $K_2{}^0$ values determined in the section 3.2 and the present one [4-8,12,14].

3.4. HSAB principle [15,16]

According to Pearson, the HSAB classifications of some species are as follows.

As hard acids: H^+, Li^+, Na^+, K^+, Ca^{2+}, Sr^{2+} *etc.*

As borderline acids: Fe^{2+}, Co^{2+}, Ni^{2+}, Cu^{2+}, Zn^{2+}, Pb^{2+} *etc.*

As soft acids: Ag^+, Tl^+, Pd^{2+}, Cd^{2+}, Hg^{2+} *etc.*

and

As hard bases: H_2O, OH^-, F^-, $PO_4{}^{3-}$, $SO_4{}^{2-}$, Cl^-, $ClO_4{}^-$, ROH, RO^-, R_2O *etc.*

As borderline bases: N_3^-, Br^-, NO_2^-, SO_3^-, N_2 *etc.*

As soft bases: R_2S, RSH, RS^-, I^-, SCN^-, CN^- *etc.* with R = aryl or alkyl group

These species are best classified by using the following criteria. Class (a) acids (hard ones) form more stable complexes with ligands having the Y donor atoms in the order Y = N >> P > As > Sb; O >> S > Se > Te; F > Cl > Br > I [15]. On the other hand, Class (b) acids (soft ones) form in the order N << P > As > Sb; O << S < Se ~ Te; F < Cl < Br < I [15]. So, what criteria do ML^{z+} classify? What criteria do L classify? For some ions and crown ethers, these HSAB classifications are going to be examined below (see 4.2) on the basis of the $K_{MXz}{}^0$ and $K_{MLXz}{}^0$ values at z = 1 and 2.

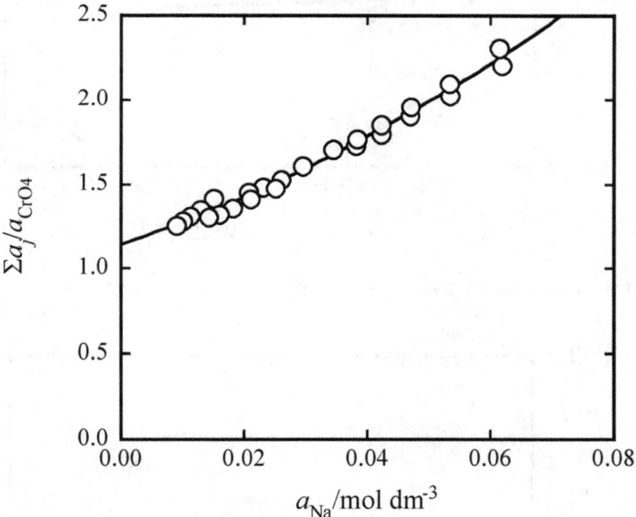

Figure 4. Plot of $\Sigma a_j/a_{CrO4}$ *versus* a_{Na} with R = 0.994

3.5. Theoretical treatments and data analysis for mixtures with L

3.5.1. For Cases (1L) and (2L) [4,8]

As similar to the section 3.2, Table 3 summarizes the analytical equations and their plot types. Examples of the plots for Cases (1L) and (2L) are shown in Figs. 5 and 6, respectively. The plot in Fig. 5 is similar to that in Fig. 3. A fitting curve of log ($K_{MLX}/y_\pm{}^2$) *versus* $a_{ML}a_X$ is depicted with a solid line in Fig. 5, where $a_{ML} = y_{ML}([X^-] - [M^+])$ [4]. The former parameter obviously corresponds to log ($K_{MX}/y_\pm{}^2$) in Case (1) and the latter one to $a_M{}^2$ {see the section 3.2.1. Table 5 lists the $K_{MLX}{}^0$ (and $\beta_2{}^0$) values thus determined at 25 °C. From this table, one can easily see that the $K_{MLX}{}^0$ values are larger than the $K_{MX}{}^0$ ones, except for several cases. These results indicate that M^{z+} dehydrates in the complex formation with L in water and thereby increases its hydrophobic property.

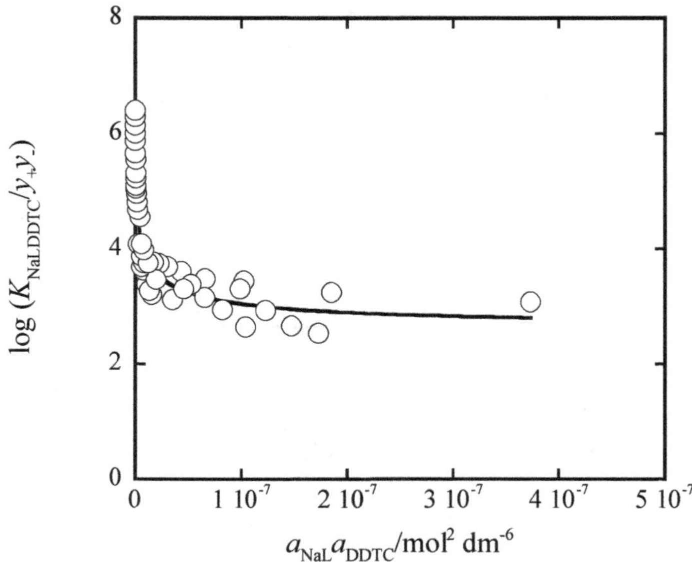

Figure 5. Plot of log ($K_{NaLDDTC}/y_{\pm}^2$) *versus* $a_{NaL}a_{DDTC}$ with R = 0.973 for L = 15C5.

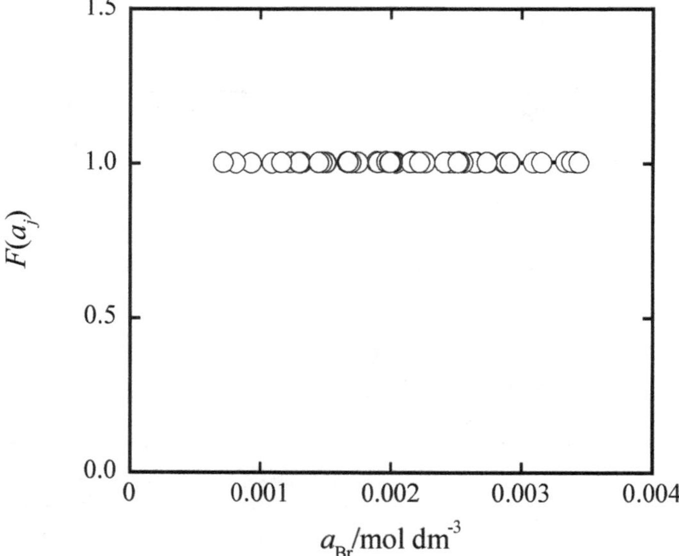

Figure 6. Plot of $F(a_j)$ *versus* a_{Br} with R = 0.484 for the $CdBr_2$-18C6 system.

4. Ion-pair formation for MX_z with L in water

4.1. Dependence of the ion-pair formation constants on some factors [4-8]

4.1.1. Effect of sizes of anions

Effect of sizes of anions X^{z-} on the $K_{MLX}{}^0$ values is described and thereby its cause is examined. Effective ionic radii or sizes estimated from the Van der Waals (vdw) volumes are on the order X^{z-} = Cl^- < Br^- < tfa^- (estimated from vdw vol.) ≤ I^- < MnO_4^- < ReO_4^- < $DDTC^-$ (from vdw vol.) < (SO_4^{2-} <) Pic^- < BPh_4^- [8]. The $K_{MLX}{}^0$ values in Table 5 are on the orders

$$X^- = Pic^- \leq MnO_4^- < BPh_4^- \text{ for } Na(15C5)^+ \text{ (O1)}$$

$$MnO_4^- < Pic^- < BPh_4^- \text{ for } Na(B15C5)^+ \text{ (O2)}$$

and

$$DDTC^- < Pic^- < MnO_4^- < ReO_4^- \leq tfa^- < BPh_4^- \text{ for } Na(18C6)^+ \text{ (O3)}$$

$$DDTC^- < ReO_4^- \leq tfa^- < Pic^- < MnO_4^- < BPh_4^- \text{ for } Na(B18C6)^+. \text{ (O4)}$$

The two orders, (O1) and (O2), suggest hydrophobic interactions between BPh_4^- and $Na(15C5)^+$ or $Na(B15C5)^+$ and those between Pic^- and $Na(B15C5)^+$. That is, the former two cases reflect the interaction between the large ions, while the latter case does that between a benzene ring in Pic^- and a benzo group in $Na(B15C5)^+$. Other $K_{MLX}{}^0$ orders seem to have the characteristics that $DDTC^-$ << BPh_4^- (is due to the hydrophobic interaction), Pic^- < MnO_4^-, and ReO_4^- ≤ tfa^- (are due to the electrostatic one).

4.1.2. Effect of a benzo group added to L skeleton

Effect of a benzo group of L on the $K_{MLX}{}^0$ values is described and thereby its cause is discussed. Table 5 reveals relations

$$L = 15C5 \leq B15C5 \text{ for NaPic, } NaMnO_4, NaBPh_4, \text{ and AgPic (R5)}$$

$$15C5 > B15C5 \text{ for KPic and } KMnO_4 \text{ (R6)}$$

and

$$18C6 < B18C6 \text{ for NaPic, KPic, } NaMnO_4, \text{ and NaDDTC (R7)}$$

$$18C6 \geq B18C6 \text{ for } KMnO_4, \text{ Natfa, } NaReO_4, NaBPh_4, \text{ and AgPic. (R8)}$$

Considering the electrostatic interaction between ML^+ and X^- and water molecules hydrated to M^+ to be basic interactions, these relations of Inequalities (R5) to (R8) may be changed into the following expression: Na^+(15C5)- ≤ (B15C5)Na^+-X^- and wAg^+(15C5)- ≤ (B15C5)Ag^+-Pic^- for Inequality (R5); (15C5)K^+w- > wK^+(B15C5)-X^- for (R6) and wNa^+(18C6)- < (B18C6)Na^+-X^-

and $K^+(18C6)$- < $(B18C6)K^+$-Pic^- for (R7); $(18C6)K^+$- ≥ $K^+(B18C6)$-MnO_4^- and $(18C6)M^+w$- ≥ $M^+(B18C6)$-X^- for (R8). Here, w denotes the water molecules which hydrate to M^+ and act as Lewis base. Also, we simply define the following sequence

$$wM^+L\text{-}X^- < M^+L\text{-}X^- < LM^+w\text{-}X^- < LM^+\text{-}X^- \text{ (SO9)}$$

as a measure for the strength of the interaction between ML^{z+} and X^{z-} at $z = 1$. In other words, the standard order (SO9) can be interpreted as L-separated ion pair with water molecule(s) < L-separated one < w-shared one < contact one. When a cavity size of L is smaller than a size of M^+, we will assume an opposite relation of LM^+w-X^- < M^+L-X^-; namely, M^+L-X^- approaches to LM^+-X^-. The relations, (R5) and (R7), seem to reflect the hydrophobic properties of ML^+. The others can reflect simply effects of the sizes of the L skeletons with benzo groups.

4.1.3. *Effect of ring sizes of L*

Effect of the ring sizes of L in MLX is described and thereby its cause is examined. Also, this means an increase in the number of the O donor atoms in the ring. These K_{MLX}^0 relations in Table 5 are

$$L = 15C5 < 18C6 \text{ and } B15C5 < B18C6 \text{ for NaPic, KPic, } NaMnO_4\text{, and } NaBPh_4 \text{ (R10)}$$

$$15C5 > 18C6 \text{ and } B15C5 > B18C6 \text{ for } KMnO_4 \text{ and AgPic. (R11)}$$

Inequality (R10) can be interpreted as the following interactions of ML^+ with X^-: $Na^+(15C5)$- < $(18C6)Na^+w$-Pic^-; $wK^+(15C5)$- < $(18C6)K^+$-Pic^-; $Na^+(15C5)$- < $(18C6)K^+w$-X^- at $X^- = MnO_4^-$ and BPh_4^-. Similarly, interactions for (R10) may be $Na^+(B15C5)$- < $(B18C6)Na^+$-X^- at $X^- = Pic^-$ and MnO_4^-; $wK^+(B15C5)$- < $(B18C6)K^+$-Pic^-. On the other hand, Inequality (R11) seems to simply reflect a coulombic interaction between ML^+ with X^-.

4.1.4. *Effect of sizes of the central M^{z+} in/on L rings*

Effective ionic radii of M^+ are on the order $M^+ = Li^+$ (0.76 Å) < Na^+ (1.02) < Ag^+ (1.15) < K^+ (1.38) [17]. Also, the hydration free energies ($-\Delta G_h^0$) of M^+ are on the order K^+ (304 kJ mol^{-1}) < Na^+ (375) < Li^+ (481) << Ag^+ (1856) [18]. The K_{MLX}^0 orders are

$$M = K < Na < Ag \text{ for M(15C5)Pic (O12)}$$

$$K < Li < Na < Ag \text{ for M(B15C5)Pic (O13)}$$

$$Li \le Na < Ag < K \text{ for M(18C6)Pic (O14)}$$

$$Ag < Na < K \text{ for M(B18C6)Pic (O15)}$$

and

$$Na < K \text{ for } M(15C5)MnO_4 \text{ (R16)}$$

K < Na for M(B15C5)MnO_4, M(18C6)MnO_4, and M(B18C6)MnO_4. (R17)

If the coulombic force is simply effective for these ion-pair formation, then the $K_{MLX}{}^0$ order can be $K^+ < Ag^+ < Na^+ < Li^+$. Similarly, if the dehydration of M^+ is effective for the formation, then the order can be $Ag^+ \ll Li^+ < Na^+ < K^+$. Comparison of these orders with (O12) to (O14) shows the complications of their ion-pair formation.

The orders or relations can be changed into wK^+(15C5)- & (15C5)K^+w- < Na^+(15C5)- < (15C5)Ag^+w- & wAg^+(15C5)-Pic^- for Inequality (O12); wK^+(B15C5)- < Li^+(B15C5)- < Na^+(B15C5)- & (B15C5)Na^+- < (B15C5)Ag^+-Pic^- for (O13); wLi^+(18C6)- ≤ wNa^+(18C6)- & (18C6)Na^+w- < (18C6)Ag^+w- & wAg^+(18C6)- < (18C6)K^+- & K^+(18C6)-Pic^- for (O14); Ag^+(B18C6)- < (B18C6)Na^+- < K^+(B18C6)-Pic^- for (O15) and Na^+(15C5)- < (15C5)K^+w-MnO_4^- for (R16); wK^+(B15C5)- & (B15C5)K^+w- < Na^+(B15C5)- & (B15C5)Na^+-MnO_4^-; K^+(18C6)- and (18C6)K^+- < (18C6)Na^+w- & wNa^+(18C6)-MnO_4^-; K^+(B18C6)- < (B18C6)Na^+-MnO_4^- for (R17). The above expression "M^+L-X^- & LM^+-X^-" means that the fraction of the left M^+L-X^- is major in comparison with that of the right LM^+-X^-.

In (O12), the inverse between $K(15C5)^+$ and $Na(15C5)^+$ is due to the fact that $K(15C5)^+$ satisfies the condition that the cavity size of 15C5 < the size of K^+ (see 4.1.2); that between $Na(15C5)^+$ and $Ag(15C5)^+$ suggests that a fraction of (15C5)Ag^+w-Pic^- is dominant. Further, the $Na(B15C5)^+ < Ag(B15C5)^+$ relation in (O13) suggests that Na^+ in the former complex ion is more shielded by B15C5 than Ag^+ in the latter ion. The same can be true of the relation of $Na(18C6)^+ < Ag(18C6)^+$ in (O14).

In the present section 4.1, the orders in magnitude among the $K_{MLX}{}^0$ values are interpreted by supposing the shapes of the MLX ion pairs based on Inequality (SO9). Of course, validity of such interpretations has to be clarified experimentally.

4.2. Try to understand the ion-pair formation based on the HSAB principle [8]

According to the HSAB classification, Pic^-, MnO_4^-, ReO_4^-, tfa^-, and BPh_4^- were reported to be hard bases [8]. Also, $DDTC^-$ has been classified as a soft base [19]. Then the $K_{MLX}{}^0$ values are on the orders

X = $DDTC^-$ ≤ Pic^- < MnO_4^- < ReO_4^- ≤ tfa^- < BPh_4^- for $Na(18C6)^+$ (O3)

$DDTC^-$ < ReO_4^- ≤ tfa^- < Pic^- < MnO_4^- < BPh_4^- for $Na(B18C6)^+$ (O4)

Here, only $DDTC^-$ has S donor atoms in it, while the other X^- does O donor atoms in them. The sequence of donor atoms in X^- obviously show S ≤ O < O < O < O (or F) for (O3) and S < O < O (or F) < O < O for (O4), except for X^- = BPh_4^-. Comparing the orders with those described in 3.4, they indicate at least that $Na(18C6)^+$ and $Na(B18C6)^+$ have the higher affinities for the O donor atoms in their X^- than the S donor ones in $DDTC^-$. This fact suggests that both the NaL^+ are classified as hard acids [8].

Similarly, from the following orders, readers can see that the same is true of the $K_{CdLX}{}^0$ values for the CdX^+-L systems with X^- = Cl^-, Br^-, and I^-.

$X^- = I^- < Br^- < Cl^-$ ($\leq Pic^-$) for Cd(18C6)X^+ and Cd(B18C6)X^+ (O18)

That is, it is suggested that both the complex ions are hard acids [8], although Cd^{2+} is classified as a soft acid (see 3.4); $K_{CdX2}{}^0$ is on the order $X^- = Cl^- \leq (Pic^- \leq) Br^- < I^-$ (see Table 5). As can be seen from the section 3.4 and the above, the facts that Cl^- and Pic^- are the hard bases support this suggestion

MX_z or M_2X	$K_{MX}{}^0$ [$K_2{}^0$], (std)[a]	MLX_z	$K_{MLXz}{}^0$ {$\beta_2{}^0$}, (std)[a]
LiPic	10.9(1.8)	Li(B15C5)Pic	205(111)
		Li(18C6)Pic	52(---[b])
NaPic	4.2(0.7)	Na(15C5)Pic	34(6)
		Na(B15C5)Pic	517(91)
		Na(18C6)Pic	62(18)
		Na(B18C6)Pic	642(96)
KPic	5.8(1.8)	K(15C5)Pic	6.0(3.9)
		K(B15C5)Pic	12(---[b])
		K(18C6)Pic	738(210)
		K(B18C6)Pic	$1.37(0.12) \times 10^3$
AgPic	2.8(0.3)	Ag(15C5)Pic	556(68)
		Ag(B15C5)Pic	$1.58(0.21) \times 10^3$
		Ag(18C6)Pic	191(19)
		Ag(B18C6)Pic	157(22)
$NaMnO_4$	8.0(0.9)	Na(15C5)MnO_4	38(21)
		Na(B15C5)MnO_4	354(75)
		Na(18C6)MnO_4	231(30)
		Na(B18C6)MnO_4	$1.63(0.41) \times 10^3$
$KMnO_4$	9.1(2.3)	K(15C5)MnO_4	239(80)
		K(B15C5)MnO_4	137(110)
		K(18C6)MnO_4	93(22)
		K(B18C6)MnO_4	72(12)
Natfa	4.0(0.8)	Na(18C6)tfa	384(67)
		Na(B18C6)tfa	201(25)
$NaReO_4$	4.1(0.7)	Na(18C6)ReO_4	340(66)
		Na(B18C6)ReO_4	188(112)
NaDDTC	32.8(2.7)	Na(18C6)DDTC	48(23)
		Na(B18C6)DDTC	100(25)
$NaBPh_4$	14.3(1.5)	Na(15C5)BPh_4	$7.36(1.51) \times 10^3$
		Na(B15C5)BPh_4	$9.07(6.30) \times 10^3$
		Na(18C6)BPh_4	$2.9(2.0) \times 10^5$
		Na(B18C6)BPh_4	$1.24(1.02) \times 10^5$
$CaCl_2$ [6]	40(7), 41(3)		
$Ca(Pic)_2$ [6]	88(58)		

MX_z or M_2X	K_{MX}^0 [K_2^0], (std)[a]	MLX_z	K_{MLXz}^0 {β_2^0}, (std)[a]
$CdCl_2$	86(30), 92(4) [8.7(7.5), 13(5)]	Cd(18C6)Cl_2	3.8 × 10^4 {1.39(1.00) × 10^6}
		Cd(B18C6)Cl_2	1.7 × 10^4, 2.70(2.48) × 10^6
$CdBr_2$	118(19) [25(19)]	Cd(18C6)Br_2	0.57(0.13)
		Cd(B18C6)Br_2	1.32(0.05)
CdI_2	308(5) [40(3)]	Cd(18C6)I_2, Cd(B18C6)I_2	< 1[c]
Cd(Pic)$_2$	108(11), 107(17)	Cd(18C6)(Pic)$_2$	3.3 × 10^4 {1.77(1.62) × 10^7}
		Cd(B18C6)(Pic)$_2$	6.5 × 10^4 {2.29(0.29) × 10^7}
$CdSO_4$	221(31)	Cd(18C6)SO_4	4.38(0.68) × 10^4
		Cd(B18C6)SO_4	1.83(0.51) × 10^4
Na_2SO_4	14(0.7) [7.4(1.0)]		
NaS_2O_3	14(0.6) [3.2(0.6)]		
Na_2CrO_4	12(2) [7.9(2.8)]		

a. Standard deviation or standard errors.
b. See Appendix B in Ref. [14].
c. Estimated from experimental results in Ref. [8].

Table 5. Ion-pair formation constants of MX_z, M_2X, and MLX_z in water at $I \rightarrow 0$ mol dm^{-3} and 25 °C [4-8,12,14]

5. Topics relevant to the ion-pair formation in water

5.1. Application to solvent extraction [20]

Using the thus-determined formation constants for M^IX and M^ILX in water, overall extraction equilibrium-constants are resolved into the following component ones [21].

$$L \rightleftharpoons L_o \quad (19\text{-}1)$$

$$M^+ + X^- \rightleftharpoons MX \quad (19\text{-}2)$$

$$M^+ + L \rightleftharpoons ML^+ \quad (19\text{-}3)$$

$$ML^+ + X^- \rightleftharpoons MLX \quad (19\text{-}4)$$

$$MLX \rightleftharpoons MLX_o \quad (19\text{-}5)$$

and

$$MLX_o \rightleftharpoons ML^+_o + X^-_o \quad (19\text{-}6)$$

Here, the subscript "o" denotes an organic phase. The reactions (19-1) to (19-5) correspond to the extraction into low polar diluents (or organic solvents) and the reaction (19-6) will be added to the cases of the extraction into high polar diluents. In general, the equilibrium constants of the process (19-1) and the reaction (19-3) can be determined by separate experiments. Then, the overall extraction equilibria are characterized on the basis of their component ones, either (19-1) to (19-5) or to (19-6). One can clearly see that, for the above extraction model, the ion-pair formation, (19-2) and (19-4), is important. The $NaMnO_4$ extraction by B15C5 into 1,2-dichloroethane (DCE) and nitrobenzene (NB) are analyzed as the results in Table 6. There are remarked differences in $K_{D,NaLX}$, $K_{D,X}$, and K_{NaLX}^{org} between DCE and NB. According to the relation $K_{ex} = K_{ML}K_{MLX}K_{D,MLX}/K_{D,L}$ [21], a difference in K_{ex} between NB and DCE mainly comes from that between $K_{D,MLX}$. Similarly, that between $K_{ex\pm}$ does from the difference between K_{MLX}^{org} in addition to $K_{D,MLX}$, because $K_{ex\pm} = K_{ex}/K_{MLX}^{org}$.

5.2. Development into study of the abnormal potential responses of ISEs [9]

Abnormal potential responses, Λ-shaped potential ones [9,12] of emf-*versus*-log $[CdX_2]_t$ plots (X = I, Br), of the commercial Cd^{2+}-selective electrode listed in Table 1 have been observed and then an answer is described using a model for the potential response with interactions of the electrode surface with counter X^-, in addition to that with Cd^{2+}. Its processes are

$$Cd^{2+} + Y^{2-}{}_s \rightleftharpoons CdY_s \text{ (subscript "s": solid phase)} \tag{20-1}$$

$$CdY_s + 2X^- \rightleftharpoons X_2CdY_s \tag{20-2}$$

$$Cd^{2+} + X^- \rightleftharpoons CdX^+, \tag{20-3}$$

where $Y^{2-}{}_s$ is a functional group of the electrode material and X^- denotes halide ions. Applying electrochemical potentials [2] for the electrode processes (20-1) and (20-2) and introducing mass-balance equations in the overall process, we obtain the following equation as an expression of emf.

$$\text{emf} = A + B\log{[Cd]_t} + C\log{\{1 + 4K_s([Cd]_t)^2\}} \tag{21}$$

Here, A, $[Cd]_t$, B, C, and K_s refer to a constant (V *versus* Ag|AgCl electrode), the total concentration of CdX_2 in the test solution, values (V) corresponding to $-2.3RT/2F$, and value ($mol^{-3}\ dm^9$ unit) being inversely proportional to the solubility product of CdX_2, respectively. One can immediately obtain these values, analyzing the plot of emf *versus* log $[Cd]_t$ by non-linear regression: $K_s(CdCl_2)$ (not be determined) < $K_s(CdBr_2)$ (= $10^{4.2}$ $mol^{-3}\ dm^9$) < $K_s(CdI_2)$ (= $10^{6.98}$) [8]. It is obvious that the larger K_s is, the more easy CdX_2 interacts with the electrode and accordingly the larger the interference of X^- against the electrode response becomes. Similar tendencies have been obtained in a commercial Cu^{2+}-selective electrode with a solid membrane and a Ca^{2+}-selective one with a liquid membrane.

Diluent	log K_{ex}[a]	log $K_{ex\pm}$[b]	log K_{ML}[c]	log K_{MLX} (I)[d]	log K_{MX}	log $K_{D,MLX}$[e]	log $K_{D,L}$[f]	log $K_{D,X}$[g]	log K_{MLX}^{org}[h] (I_o)[i]
DCE	2.24	-3.7_5	0.45	2.47 (0.0084)	0.54	1.23	1.910	−3.3	6.0 (4.5×10^{-6})
NB	3.79	-0.2_3	0.45	2.51 (0.0015)	0.59	2.4	1.6	−1.7	4.0 (3.5×10^{-5})

a. Defined as $K_{ex} = [MLX]_o/[M^+][L]_o[X^-]$. b. Defined as $K_{ex\pm} = [ML^+]_o[X^-]_o/[M^+][L]_o[X^-]$. c. As $K_{ML} = [ML^+]/[M^+][L]$. d. Ionic strength in a water phase. e. As $K_{D,MLX} = [MLX]_o/[MLX]$. f. As $K_{D,L} = [L]_o/[L]$. g. As $K_{D,X} = [X^-]_o/[X^-]$. h. Estimated from $K_{MLX}^{org} = K_{ex}/K_{ex\pm}$. i. Ionic strength in the o phase.

Table 6. Extraction constants and their component equilibrium-constants for the $NaMnO_4$-B15C5 extraction systems into DCE and NB at 25 °C [20]

6. Summary

In this chapter, the determination procedures of the $K_{MXz}{}^0$, $K_{M2X}{}^0$, and $K_{MLXz}{}^0$ values in water were discussed and their some applications were described. The present potentiometric procedures will support other methods, such as conductometric, spectrophotometric, solvent extraction ones, and so on. Their applications have been limited to some ions and hydrophilic L, because kinds of commercial ISE is limited and low solubility of hydrophobic L to water limits its emf measurements; the solvent extraction method [21] is going to compensate for the latter limitation. Instead of such situations, the present procedures are useful for the $K_{MLXz}{}^0$ determination. As the pH values are measured at many laboratories in the world, thus handy procedures, compared with the other methods, will take effect at the determination of K_{MXz} and K_{MLXz} at $I \to 0$.

Author details

Yoshihiro Kudo
Chiba University, Japan

Appendix

Symbols and abbreviations used in this chapter

Symbol and abbreviation	Meaning
$a(j)$	Ion size parameter of j
B15C5	Benzo-15-crown-5 ether
B18C6	Benzo-18-crown-6 ether
c_j^{I} or c_j^{II}	Molar concentration of j in the phase I or II
$DDTC^-$	Diethyldithiocarbamate ion
DME	Dropping mercury electrode
Ej	Liquid junction potential at an interface
$I \to 0$	Ionic strength at infinite dilution
$[j]$	Molar concentration of j at equilibrium
$[j]_t$	Total concentration of j

K	The equilibrium constant
Pic^-	Picrate ion
R	Correlation coefficient
RT/F	0.02569 V at 298 K
tfa	Trifluoroacetate ion
uj	Mobility of *j* in water
z	Formal charge of an ion or a number expressing the composition of a salt or an ion pair
zj	Formal charge of *j*
ϕ^I or ϕ^{II}	Inner potential in the phase I or II
2.3*RT/F*	0.05916 V at 298 K

7. References

[1] G. D. Christian, "*Analytical Chemistry*", 5th ed., John Wiley & Sons, New York, 1994.

[2] A. J. Bard and L. R. Faulkner, "*Electrochemical Methods: Fundamentals and Applications*", 2nd ed., John Wiley & Sons, New York, 2001.

[3] S. Katsuta, H. Wakabayashi, M. Tamaru, Y. Kudo, and Y. Takeda, *J. Solution Chem.*, 2007, *36*, 531.

[4] Y. Kudo, M. Wakasa, T. Ito, J. Usami, S. Katsuta, and Y. Takeda, *Anal. Bioanal. Chem.*, 2005, *381*, 456.

[5] Y. Kudo, R. Fijihara, T. Ohtake, M. Wakasa, S. Katsuta, and Y. Takeda, *J. Chem. Eng. Data*, 2006, *51*, 604.

[6] Y. Kudo, S. Takeuchi, Y. Kobayashi, S. Katsuta, and Y. Takeda, *ibid.*, 2007, *52*, 1747.

[7] Y. Kudo, Y. Kobayashi, S. Katsuta, and Y. Takeda, *J. Mol. Liquids*, 2009, *146*, 60.

[8] Y. Kudo, T. Koide, Y. Zhao, S. Katsuta, and Y. Takeda, *Anal. Sci.*, 2011, *27*, 1207.

[9] Y. Kudo, D. Todoroki, N. Suzuki, N. Horiuchi, S. Katsuta, and Y. Takeda, *Amer. J. Anal. Chem.*, 2011, *2*, 9.

[10] J. Kielland, *J. Am. Chem. Soc.*, 1937, *59*, 1675.

[11] P. Vanýsek, "*Ionic Conductivity and Diffusion at Infinite Dilution*" in "*CRC Hand Book of Chemistry and Physics*", ed. D. R. Lide, 74th ed., CRC Press, Boca Raton, 1993.

[12] Y. Kudo, D. Todoroki, N. Horiuchi, S. Katsuta, and Y. Takeda, *J. Chem. Eng. Data*, 2010, *55*, 2463,

[13] R. de Levie, "*Aqueous Acid-Base Equilibria and Titration*", ed. R. G. Compton, Oxford Chemistry Primers 80, Oxford University Press, New York, 1999.

[14] Y. Kudo, J. Usami, S. Katsuta, and Y. Takeda, *J. Mol. Liquids*, 2006, *123*, 29.

[15] R. G. Pearson, *J. Chem. Educ.*, 1968, *45*, 581.

[16] R. G. Pearson, *Inorg. Chim. Acta*, 1995, *240*, 93.

[17] R. D. Shannon, *Acta Crystallogr.*, 1976, *A32*, 751.

[18] Y. Marcus, "*Ion Properties*", Marcel Dekker Inc., New York, 1997.

[19] H. Kawamoto and H. Akaiwa, *Chem. Lett.*, 1990, 1451.

[20] Y. Kudo, K. Harashima, S. Katsuta, and Y. Takeda, *International J. Chem.*, 2011, *3*, 99.
Y. Takeda, R. Taguchi, and S. Katsuta, *J. Mol. Liquids*, 2004, *115*, 139.

Chapter 6

Electroanalytical Sensor Technology

Aoife C. Power and Aoife Morrin

Additional information is available at the end of the chapter

1. Introduction

The subject of electrochemical sensors is broad, spanning many aspects of physical and analytical chemistry, materials science, biochemistry, solid-state physics, device fabrication, electrical engineering, and even statistical analysis. Thus, the field of electrochemical sensors cannot be dealt with holistically in a single chapter. Here, we will focus on electrochemical sensor technology from an analytical perspective, where the rigours of sensor behaviour will be discussed as they relate to the quality of the quantitative information that can be derived. The definition of analytical chemistry was given by the Federation of European Chemical Societies (FECS) in 1993 [1] and adopted by IUPAC:

> "*develops and applies methods, instruments and strategies to obtain information on the composition and nature of matter in space and time*, as well as on the value of these measurements, i.e., their uncertainty, validation, and/or traceability to fundamental standards."

Electroanalytical chemistry, also known as electroanalysis, lies at the interface between analytical science and electrochemistry. It is concerned with the development, characterisation and application of chemical analysis methods employing electrochemical phenomena. It has major significance in modern analytical science, enabling measurements of the smallest chemical species, right up to the macromolecules of importance in modern biology. The history of electrochemical sensors starts basically with the development of the glass electrode by Cremer in 1906 [2]. Haber and his student Klemensiewicz took up the idea in 1909 and made the basis for analytical applications [3]. The former wanted to introduce the device as "Haber electrode" causing protests of Cremer. The latter should be given full appreciation of the invention of the glass electrode though Haber dominates the literature [4]. Today, the electrochemical sensor plays an essential analytical role in the fields of environmental conservation and monitoring, disaster and disease prevention, and industrial analysis. A typical chemical sensor is a device that transforms chemical information in a selective and reversible way, ranging from the concentration of a specific sample component to total composition analysis, into an analytically useful signal. A huge research effort has taken place over the several years to achieve electrochemical sensors with attractive qualities including rapid response, low cost, miniaturisable, superior sensitivity and selectivity, and appropriate detection limits. Approximately 2000 peer-reviewed papers concerning electrochemical sensors were published in 2011 according to Thomson Reuters Web of Knowledge® showing the considerable research effort underway in this field.

In the highly diverse field of chemical (and biochemical) sensing, the sensor is governed by both the aspect of the environment it is measuring and the matrix in which it is in. As well as sensors that use electrochemistry as the type of energy transfer that they detect, optical [5], thermal [6] and mass-based [7] sensors are also well-developed. From an analytical perspective, electrochemistry is appealing as it directly converts chemical information into an electrical signal with remarkable detectability, experimental simplicity and low cost. There is no need for sophisticated instrumentation, e.g., optics. A very attractive feature of electrochemistry is that it depends on a surface phenomenon, not an optical path length, and thus sample volumes can be very small, lending itself to miniaturisation.

The interest in electrochemical sensors continues unabated today, stimulated by the wide range of potential applications. Their impact is most clearly illustrated in the widespread use of electrochemical sensors seen in daily life, where they continue to meet the expanding need for rapid, simple and economic methods of determination of numerous analytes [8-10]. Through the refinement of existing sensing technologies along with the development of innovative functional sensor materials including nano- and biological materials [11-14], improved data analysis [15], and sensor fabrication and miniaturisation [16-18], opportunities for the construction of new generation sensors with much improved performances are emerging. Two branches of electrochemical sensors are developing: sensors with increased specificity and sensors capable of simultaneous/multiplex determination. In both of these branches, the ability to operate in complex biological matrixes will remain critical, forcing researchers to solve problems of biocompatibility and stability [19]. As such, the analytical and physical properties that must be considered when developing (and commercialising) chemical and biological sensors include (but are not limited to):

- Cost
- Miniaturisation
- Sensitivity
- Sensor reproducibility
- Selectivity/Specificity
- Multi-analyte detection
- Stability

Few sensors, if any, exhibit optimal characteristics for all properties. For example, for *in vivo* application, due to the complex environment of these measurements, sensing selectivity becomes critical and a trade-off between selectivity and other parameters such as cost or response time is needed [20]. When developing electrochemical sensors, some properties can be prioritised over others depending on final application. An electrocatalytic hydrogen peroxide sensor with a low milli-molar (mM) detection range may, for example, have application in waste water or industrial applications. However, in order to be viable for application in clinical diagnostics, it would demand a sensitivity several orders of magnitude lower.

Following an overview of the standard sensing technologies and a brief introduction to biosensors, this chapter will review chemical and biological sensors under the criteria listed above, discussing the latest research developments in these areas published in the peer-review literature in the last three years. We will focus on all the facets of electroanalytical sensing technology with particular emphasis on the impact of nanotechnology and nanomaterials, microfabrication and biotechnology on the field to date.

2. Electrochemical sensing principles

Depending on the exact mode of signal transduction, electrochemical sensors can use a range of modes of detection such as potentiometric, voltammetric and conductimetric. Each principle requires a specific design of the electrochemical cell. Potentiometric sensors are very attractive for field operations because of their high selectivity, simplicity and low cost. They are, however, less sensitive and often slower than their voltammetric counterparts. In the past, potentiometric devices have been most widely used, but there is an increasing amount of research being carried out on amperometric sensors that has tipped this balance. There are also sensors relying simply on conductivity changes of ions, but there is a far larger group of sensors, which work on resistivity and impedance, such as chemiresistors and capacitive sensors. As the underlying operating principle for conductimetric sensors is usually not an electrochemical reaction or property, they should be classified as electrical chemical sensors rather than as electrochemical ones. While most electrochemical techniques for sensing analytes of interest are based on the changes in potential or current, Shan *et al.* [21] have developed a completely novel method for performing electrochemical measurements. In their work, they report a method for imaging local electrochemical current using the optical signal of the electrode surface generated from a surface plasmon resonance (SPR). The electrochemical current image is based on the fact that the current density can be easily calculated from the local SPR signal. The authors demonstrated this concept by imaging traces of TNT on a fingerprint on a gold substrate.

More detailed theoretical discussions on potentiometric, voltammetric and conductimetric measurements are available in this and many other textbooks [22-24] and so will not be discussed in detail here.

2.1. Potentiometric sensors

In potentiometric sensors, the analytical information is obtained by converting the recognition process into a potential signal, which is logarithmically proportional to the concentration (activity) of species generated or consumed in the recognition event. The Nernst equation logarithmically relates the measured electrode potential, E, to the relative activities of the redox species of interest:

$$E = E^O + \frac{RT}{nF}\ln\frac{a^O}{a^R} \tag{1}$$

Where $E^O$is the standard electrode potential (quoted relative to the idealised Standard Hydrogen Electrode, SHE) and $a^O$and a^R are the activities of the oxidised and reduced species, R is the universal gas constant; T is the absolute temperature; F is the Faraday constant; n is the number of moles of electrons exchanged in the electrochemical reaction.

The most representative potentiometric sensor is the ion selective electrode (ISE)[25]. The ISE uses an indicator electrode which selectively measures the activity of a particular analyte ion. An ion-selective membrane, placed at the tip of the electrode, is designed to yield a potential signal that is selective for the target ion. This potential signal is generated by a charge separation at the interface between the ion-selective membrane and the solution due to selective partitioning of the ionic species between these two phases. The response is measured under conditions of essentially zero current. The response of the indicator electrode should be fast, reversible and governed by the Nernst equation.

In classical ISEs, the arrangement is symmetrical [26] which means that the membrane separates two solutions, the test solution and the inner solution with a constant concentration of ionic species. The electrical contact to an ISE is provided by a reference electrode (usually Ag/AgCl) in contact with the internal solution that contains chloride ions at constant concentration. The measured ISE potential is the sum of the two reference electrode potentials, the membrane potential constituted by boundary potentials at each membrane/solution interface, and a possible diffusion potential which may be caused by an ion concentration gradient within the ion-selective membrane phase.

ISEs with solid inner contacts are considered to be asymmetrical [26]. Taking into consideration that potentials generated at each membrane interface are included in the overall sensor signal response it is clear that to obtain a solid-contact ISE with a stable electrode potential, a fast and thermodynamically reversible ion-to-electron transduction in the solid state is required without any contribution from parasitic side reactions. Solid-contact potentiometric ion-selective electrodes have nowadays similar performance characteristics to conventional inner-solution ion-selective electrodes and offer new and advantageous technical possibilities such as miniaturization to the μm scale, cost-effective fabrication, no need for maintenance, flexibility, and multiple shape configurations. Among the electroactive materials available today, conducting polymers have emerged as a promising ion-to-electron transducers for solid-contact ISEs [25, 27, 28]. In this type of solid-contact ISE, the conducting polymer is coated with a conventional ion-selective membrane, and the ion-selectivity is determined mainly by the ion-selective membrane. For example, potentiometric sensors based on a glassy carbon electrode covered with polyaniline and various thiacalix[4]arene ionophores have been developed and applied for the successful determination of Ag^+ ions [29]. Other configurations for solid-contact ISEs recently reported involve the use of carbon cloth coated with a conventional plasticized PVC-based K(+)-selective membrane [30], carbon nanotubes (CNTs) drop-coated with a K^+-selective polyacrylic membrane [31], PVC-based molecularly imprinted polymers [32] and graphene coated with an ionophore-doped polymeric membrane [33].

ISEs are been combined with semiconductor field effect transistor (FET) technology to give ion-selective FETs, where the gate has been replaced by an ion-selective membrane. ISFETs with bare gate insulator (silicon oxide, silicon nitride, aluminium oxide, etc.) show intrinsic pH-sensitivity due to electrochemical equilibrium between protonated oxide surface and protons in the solution. To obtain sensitivity to other ions a polymeric membrane containing some ionophore is deposited. The advantages of ISFETs include small size and rugged construction, making it a useful sensing technology for environmental and industrial analysis.

Light addressable potentiometric sensors (LAPS) [34, 35] is another type of field-effect transducer that are used as potentiometric chemical sensors which has gained reasonable popularity. Their principle of operation is quite similar to that of ISFETs in which the drain-source current in a space charge region at the semiconductor/insulator interface depends on the applied gate potential. Illumination by light source with modulated intensity generates an AC photocurrent that depends on the applied potential. Like ISFET sensors, ion-selective membranes of various types may be deposited onto the insulator surface to give the required ion sensitivity [26].

2.2. Voltammetric sensors

Voltammetry provides an electroanalytical method, the premise of which is that current is linearly dependent upon the concentration of the electroactive species (analyte) involved in a chemical or biological recognition process (at a scanned or fixed potential). Voltammetry implies a varying voltage. Cyclic voltammetry, squarewave and stripping voltammetry are some of the more common techniques. Amperometry is strictly a sub-class of voltammetry in which the electrode is held at constant potentials for various lengths of time. The distinction between amperometry and voltammetry is mostly historic as there was a time when it was difficult to switch between "holding" and "scanning" a potential. This function is trivial for modern potentiostats, and today there is little distinction between the techniques which either "hold", "scan", or do both during a single experiment. The recent review by Gupta *et al.* [36] focuses on the various applications of voltammetry to pharmaceutical analysis from 2001 to 2010.

Stripping analysis is one of the most sensitive voltammetric methods [37, 38]. Such techniques enjoy the advantages that there is no need for derivatization and that these methods are less sensitive to matrix effects than other analytical techniques. It is mostly used for trace analysis of heavy metals for environmental analysis [39, 40]. Cathodic or anodic stripping voltammetry have also been used for a highly sensitive determination of nucleobases, nucleosides, nucleotides or acid-hydrolyzed NAs, based on formation of sparingly soluble complexes of the NA constituents with electrochemically generated mercury or copper(I) ions [41].

In amperometry, the working electrode is held at a constant potential while the current is monitored. The current is then related to the concentration of the analyte present. This sensing method is commonly employed in both biosensors and immunosensors, which will be discussed further later. The first amperometric sensor was the oxygen electrode developed by L.C. Clark [42]. Oxygen entering the system through a gas-permeable membrane is reduced to water at a noble metal cathode. Clark also described the first glucose biosensors in 1962, using his oxygen electrode to determine the depletion of oxygen by the action of glucose oxidase on glucose [43]. Today, about 50 years on, Hu *et al.* [44] reported an oxygen sensor based on an inkjet printed nanoporous gold electrode array on cellulose membranes using ionic liquid as electrolyte. The sensor looks like a piece of paper but possesses high sensitivity for O_2 in a linear range from 0.054 to 0.177 v/v %. Hydrogen peroxide sensors as well as non-enzymatic glucose sensors are of particular interest in the field of amperometric sensors [20].

The development of all voltammetric techniques has been based predominantly on the use of mercury, carbonaceous materials and noble metals as working electrodes. However, toxicity of mercury, inconvenience in working with liquid hanging drop electrodes, and a limited range of potentials for mercury for anodic reactions has essentially eliminated mercury from this list. Carbon–based working electrode materials include all allotropic forms of carbons - graphite, glassy carbon, amorphous carbon, fullerenes, nanotubes, and are all used as important electrode materials in electroanalytical chemistry. Over the past five decades, carbon paste, i.e., a mixture of carbon (graphite) powder and a binder (pasting liquid), has arguably become one of the most popular carbon electrode materials used for sensors given that carbon pastes can readily be screen-printed for the mass-production of electrodes. Screen-printing is a thick-film process, which has been used for many years in artistic applications and, more recently, for the production of miniature, robust and cheap electronic circuits. During the early 1980s, the process was adapted for the production of carbon-paste-based amperometric biosensors which had a huge impact on biosensor commercialisation. The majority of most successful electrochemical sensors, including the blood glucose biosensor strip, to date, employ a screen-printed carbon-paste as the working electrode [45, 46].

The use of new materials, especially nanomaterials, has become an increased area of research in electrochemical sensors. The incorporation of these nanomaterials in conjunction with one another to form novel composites is particularly interesting, as many of these materials have been found to have synergistic effects. Zhong *et al.* [47] reported a non-enzymatic hydrogen peroxide amperometric sensor based on a glassy carbon electrode modified with an MWCNT/polyaniline composite film and platinum nanoparticles. The composite films have a large specific surface, high redox electrochemical activity and good environmental stability. Due to the synergistic catalytic activity between MWCNTs-PANI nanocomposites and Pt nanoparticles, the fabricated non-enzyme amperometric sensor displayed a high sensitive detection of H_2O_2. CNTs were also used by Guo *et al.* [48] who developed a sensitive amperometric sensor for tryptophan by modifying a glassy carbon (GC) electrode with gold nanoparticle decorated CNTs. The nanocomposite material demonstrated synergistic enhancement for the electrocatlytic activity toward the oxidation of tryptophan.

2.3. Conductimetric and impedimetric sensors

Conductimetric sensors are based on the measurement of electrolyte conductivity, which varies when the cell is exposed to different environments. The sensing effect is based on the change of the number of mobile charge carriers in the electrolyte. If the electrodes are prevented from polarizing, the electrolyte shows ohmic behaviour. Conductivity measurements are generally performed with an AC supply. The conductivity is a linear function of the ion concentration; therefore, it can be used for sensor applications. Ventura *et al.* [49] have reported a paper-based conductivity sensor for hydrogen. Palladium nanoparticles were attached onto a covalently cross-linked CNT paper. Depending on the design of the sensor, p-type or n-type conduction was observed as the resistance of the paper was measured under exposure to hydrogen gas. This paper demonstrates a simply designed, flexible gas sensor, highlighting the next stage in the development of functional nanotube paper. Another format reported was an electrospun nylon 6,6 nanofiber containing an ionic liquid (1-butyl-3-methylimidazolium hexafluorophosphate) was electrospun and found to behave as an effective, reversible chemiresistor for detecting organic vapours, highlighting another route to a simple, flexible gas conductivity sensor [50]. Conductivity measurements in general, are non-specific for a given ion type. Also, the polarization and the limiting current operation mode need to be avoided as these can damage the electrode interface. Thus, a small amplitude alternating bias can be used for sensor measurements with frequencies where the capacitive coupling is still not determining the impedance measurement. This was used for the sensing of ammonia on polyaniline-modified interdigitated arrays to prevent polarisation of the polyaniline [51].

Impedance-based sensors have a similar design as mixed potential type sensors. Instead of measuring the voltage, a sinusoidal voltage is applied and the resulting current is measured. Impedance is then calculated as the ratio of voltage to current in the frequency domain. By using small amplitude sine wave perturbation, linearity in electrochemical systems can be assumed, allowing frequency analysis. An excellent review of the use of impedance for the sensing of gases has recently been published by Rheaume & Pisano [52].

The use of impedance/capacitance has also been used to detect the antibody/antigen complex formation which has application in electrochemical immunosensors. This phenomena was first reported in 1998 when upon formation of an Ab/Ag complex on the surface of the electrode, the increase in dielectric layer thickness caused changes in capacitance proportional to the size and the concentration of antibodies [53]. Impedance changes between electrode surfaces and a surrounding solution upon a binding event can be transduced into an electrical signal using a frequency response analyser. There are several theories as to how this binding event affects changes in real and imaginary components of the system, although it is difficult to identify the origin of these changes. One theory hypothesises that binding of larger antigens forms a resistive barrier, causing the impedance to increase whilst binding of smaller antigens can acilitate a charge transfer and lower impedance [54]. Future work must establish the origin of this impedance change, whether from increases in surface density or perhaps from conformational changes that modify charge transfer across the sensor interface. Recent examples of impedance-based immunosensors include a silicon nitride (Si_3N_4) surface with covalently bound anti-human serum albumin (anti-HSA) [55], Electrochemical impedance spectroscopy (EIS), measurements for the specific detection of HSA proteins where a detection limit of 10^{-14} M were achieved. This is one of a handful of papers that have generated EIS measurement on insulating surfaces. Mostly impedimetric immunosensors are based on conducting materials. Chemically modified graphene (CMG), immobilised on a printed electrode was demonstrated as a disposable platform for the attachment of anti-IgG (anti-Immunoglobulin) [56]. The principle of detection lies in the changes in impedance spectra of a bulk solution-based redox probe (10 mM $K_4[Fe(CN)_6]/K_3[Fe(CN)_6]$) after the attachment of IgG to the immobilized anti-IgG. The immunosensor was optimised and it was found that thermally reduced graphene oxide has the best performance in terms of sensitivity when compared to other CMG materials.

3. Biosensors

Biosensors aim to utilise the power of electrochemical techniques for biological processes by quantitatively producing an electrical signal that relates to the concentration of a biological analyte. The relatively low cost and rapid response of these sensors make them useful in a variety of fields including healthcare, environmental monitoring, and biological analysis among others [19]. Biosensors use biomolecules as recognition elements which must be immobilised (or coupled) to the transducer. The transducer is the electrode (or set of electrodes) in the case of the electrochemical sensor. The specificity and selectivity that a biosensor provides is attributed to this immobilised biological recognition element. Biocatalytic and bioaffinity recognition elements are the two classes of molecules employed. Enzymes, which are biocatalytic recognition elements are the best known and studied. Recent development has focused on improving the immobilization and stability of the enzymes [57, 58]. Enzymes immobilized to nanosized scaffolds such as spheres, fibres and tubes have all recently been reported [59-61]. The premise of using nanoscale structures for immobilization is to reduce diffusion limitations and maximize the functional surface area to increase enzyme loading. In addition, the physical characteristics of nanoparticles such as enhanced diffusion and particle mobility can impact inherent catalytic activity of attached enzymes [62]. Increased enzyme stability at these surfaces is also widely reported. A novel platform based on buckypaper (which is a thin membrane of CNT networks) was reported using glucose oxidase as the model enzyme. A biocompatible mediator-free biosensor was studied and the potential effect of the buckypaper on the stability of the biosensor was assessed. The results showed that the biosensor had a considerable functional lifetime of over 80 days [58]. Greatly enhanced electrochemistry of the enzyme can be observed by the use of nanoparticles, due to the ability of the nanoparticles to reduce the distance between the redox centre of the enzyme and the electrode. Willner's group has contributed significantly to this area by taking the approach of reconstituting apo-proteins on cofactor-modified electrodes as a general strategy to electrically contact redox enzymes with electrodes [63, 64].

A recent news article in Nature Chemistry [65] points to the major Achilles heel of enzyme-based sensors. To measure glucose, enzymes such as glucose oxidase, hexokinase, or glucose dehydrogenase can be coupled to reactions that generate a signal. But such an analytical set-up can be used only if analogous enzymes are available for the analyte of interest. They identify a paper by Lu & Xiang [66] that overcomes this severe limitation where a method (Figure 1) for expanding the principle of the glucose meter to detect other analytes is reported. Instead of measuring the analyte directly, their strategy aims to produce glucose in quantities proportional to the amount of the analyte of interest. The approach uses magnetic microbeads coated with an enzyme (invertase) that is conjugated to either DNA aptamers or DNAzymes. These DNA oligomers can be selected from huge libraries of random DNA sequences, based on their binding to an analyte of choice. When the target binds to the aptamer or DNAzyme, invertase is released into the solution. After magnetic separation of microbeads, the invertase hydrolyses sucrose into glucose, which can then be quantified using a conventional glucose meter. Lu and Xiang demonstrate the measurement of four different classes of non-glucose analytes: a small molecule (cocaine), a nucleoside (adenosine), a protein (interferon-gamma) and a metal ion (uranium).

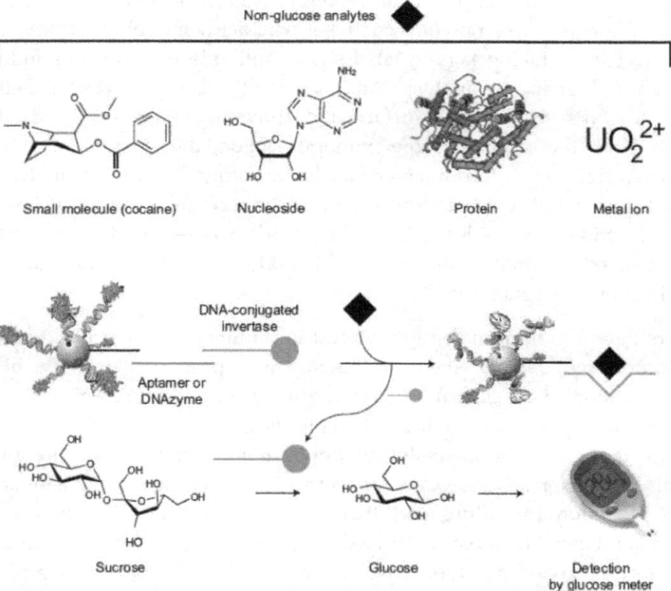

Figure 1. Sensing a variety of analytes using a glucose meter. Aptamers or DNAzymes that bind a wide range of analytes can be selected from huge libraries of random DNA sequences. When the analyte binds to its aptamer, invertase is released. Sucrose is then converted to glucose in proportion to the original analyte and can be detected using a standard glucose meter. Reprinted by permission from Macmillan Publishers Ltd: [Nature Chemistry] (Sia SK and Chin CD, Analytical Chemistry: Sweet Solution to Sensing, Nature Chemistry, 3: 659-660.), copyright (2011).

The second class of bioaffinity recognition elements such as antibodies and DNA are also widely used in biosensing. Immunosensors, which perform immunoassays based on antigen and antibody recognition, have become vital for the determination of biochemical targets relating to health concerns spanning from cancer antigens [67] to pathogens [68]. Some of the most significant advances include development of immunosensors for the continuous monitoring of analytes, point of care (PoC) devices, with lower unit costs, automation, reusability and ease of use [8]. A continual concern with immunosensor development is the capability to sensitively detect relevant immunological compounds without compromising the bioactivity of the immunoactive species on the electrode. Stabilisation of sensor constructs was achieved by using polylysine films into which anti-biotin molecules were immobilised [69]. Cross-linking prevents conformational change and unfolding of the antibodies allowing markedly enhanced sensitivity when compared with similar constructs, longer storage times and higher resistance to extremes of pH and temperature. Many researchers have increasingly utilized nanomaterials to support the immunoactive agents while simultaneously enhancing the electrochemical and analytical capabilities of the electrode [70, 71]. A triple signal amplification strategy was designed by Lin *et al.* [17] for the ultrasensitive immunosensing of a cancer biomarker, carcinoembryonic antigen (CEA). This strategy was achieved using graphene for accelerating electron transfer, poly(styrene-co-acrylic acid) microbead (PSA) carried Au nanoparticles (NPs) as tracing tags to label signal antibody and Au NPs induced silver deposition for anodic stripping analysis. An *in situ* synthesis of Au NPs led to the loading of numerous Au NPs on the PSA surface and convenient labelling of the tag to the secondary Ab. With the sandwich-type immunoreaction, the anchored Au NPs further induced the chemical deposition of silver for electrochemical stripping analysis of target antigen. The triple signal amplification greatly enhanced the sensitivity for biomarker detection. A detection limit as low as to 0.12 pg/mL was reported. The immunosensor exhibited good stability and acceptable reproducibility and accuracy, indicating potential applications in clinical diagnostics.

In recent years there has been increasing interest in finding new molecules that are able to mimic antibodies and replace them for therapeutic purposes and for bioanalytical applications. Antibodies are difficult and expensive to manufacture and are produced in vivo, by immunizing animals. Different molecules are currently being studied as alternatives to antibodies for bioanalytical applications; among these are nucleic acid aptamers. One of the main advantages of nucleic acid aptamers compared with antibodies is their in vitro selection procedure and their chemical synthesis. These manufacturing procedures do not depend on a particular analyte (possibility of using toxins, and molecules that do not elicit a good immune response) and enable the use of non-physiological conditions (including extremely high or low temperature and pH), do not require animals and cell lines, and are cost-effective. Most of the electrochemical detection principles described for other bioaffinity assays are applicable to electrochemical aptamer-based biosensors, or aptasensors. Label-free modes are based on the change in electrode surface behaviour after formation of the aptamer–target complex (usually monitored by electrochemical impedance spectroscopy [19, 72] or by use of a FET [73].

4. Physical and analytical properties of electroanalytical sensors

4.1. Cost

Electrochemical sensors provide a low cost analytical tool. Furthermore, the ability to produce electrochemical sensors in large numbers at a low cost is a major requirement for many applications. For instance, for commercial sensors aimed at the medical self-testing and smart packaging markets, single-shot use has many advantages as it immediately overcomes issues with cross-contamination and integrating sensors into disposable packaging. Thus, the ability to mass manufacture sensors, with minimal production costs facilitates the potential of sensors as low-cost disposable platforms.

The current state-of-the-art in sensor manufacturing is the commercial glucose test strip which is produced in volumes of billions annually. Given the volumes being manufactured, the cost per sensor is a fraction of a cent. The manufacturing technology that has been exploited so successfully for this is screen-printing [46]. Screen-printing is a thick-film process, which has been used for many years in artistic applications and, more recently, for the production of miniature, robust and cheap electronic circuits. During the early 1980s, the process was adapted for the production of amperometric biosensors which had a huge impact on biosensor commercialisation. The majority of most successful electrochemically based devices, including the blood glucose biosensor strip, to date, have used screen-printing as the manufacturing tool [74]. Another type of low-cost printing is also being adopted for sensor fabrication is inkjet [44, 75, 76]. It has the required levels of flexibility, resolution and scale-up required for sensor production. It remains to be seen however if it is adopted by the electrochemical sensor community. Crooks research group recently reported a paper-based electrochemical sensing platform with integral battery and electrochromic read-out [77] (Figure 2). The platform is fabricated based on paper fluidics and uses a Prussian blue spot electrodeposited on an indium-doped tin oxide thin film as the electrochromic indicator. Although this is not a printed platform, the concept of using screen- and/or inkjet-printing to mass produce such a sensor could be easily envisaged. It is likely however, that a combination of print and possibly non-print methods (e.g., photolithography) will be adopted in the future in order to produce systems comprising not only the sensor component, but also the display, power and circuitry components in order to build highly sophisticated, low-cost autonomous systems [45].

4.2. Miniaturisation

Through the reduction in size of both the functionally critical components of sensing devices and the sensors themselves, miniaturisation offers a range of distinct advantages to the analyst, including the reduction of:

- transport times and volumes
- dead volumes
- sample preparation
- reagent consumption
- energy consumption
- time expenditure and monetary cost.

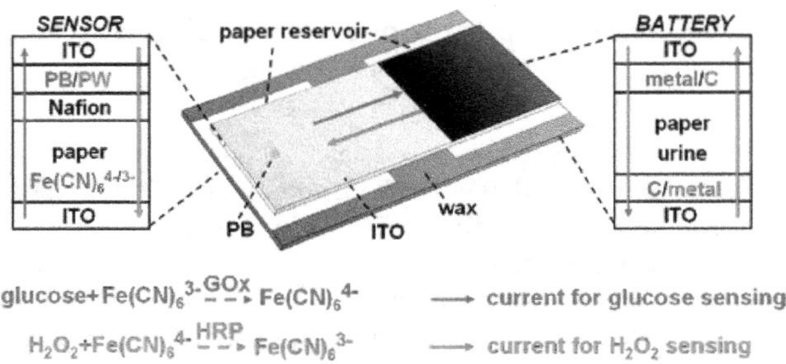

Figure 2. Drawing illustrating the operational principles of a paper-based electrochemical sensing platform with integral battery and electrochromic read-out. The device consists of two parts: a sensor section and an Al/air battery section, which are separated by a wax barrier. For glucose detection, the paper reservoir of the sensor section is preloaded with dried GOx and $Fe(CN)_6^{3-}$. The catalytic oxidation of glucose by GOx results in conversion of $Fe(CN)_6^{3-}$ to $Fe(CN)_6^{4-}$. $Fe(CN)_6^{4-}$ is then oxidized back to $Fe(CN)_6^{3-}$ on the lower ITO electrode, which results in reduction of PB to colorless PW on the upper ITO electrode. The H_2O_2 sensor operation is similar, except that HRP catalyzes the oxidation of $Fe(CN)_6^{4-}$ to $Fe(CN)_6^{3-}$ in the presence of H_2O_2. The resulting $Fe(CN)_6^{3-}$ is then reduced at the lower ITO electrode, while PW is oxidized to PB at the upper ITO electrode. The battery section of the device drives the electrochemical reactions in the sensor section [77]. Reprinted with permission from (Liu H and Crooks RM, Paper - Based Electrochemical Sensing Platform with Integral Battery and Electrochromic Read-Out, Analytical Chemistry, 84: 2528-2532). Copyright (2012) American Chemical Society.

This coupled with improved portability has resulted in miniaturisation becoming a major driving force in sensor research, prompting not only the scaling down of established sensing devices [78], but also the development and application of novel sensing materials [79, 80] in a variety of sensors. At present the majority of electrochemical sensor development is application orientated which has resulted in the extreme miniaturisation of sensing devices. This is in a large way due to the concurrent advances in nanotechnology which has allowed the fabrication of novel sensors possessing signal transduction mechanisms that exploit the unique physical phenomena of the nanoscale [81]. These sensors may utilise, for example, the spectroscopy of plasmonic nanoparticles, the deflection of cantilevers and the conductivity of nanowires [81, 82].

The focus of this section will be on the development of novel sensors produced through novel fabrication techniques, which show comparable or improved sensing performance with conventional sensors while also highlighting improvements in the ideal sensor features listed in the chapter's introduction. Success in terms of the improved selectivity and stability of miniaturised potentiometric sensors will be further discussed in following sections.

The impact miniaturisation is having on the field of electrochemical research is most readily visualised when we consider the development of biosensors, specifically those that are commercially available, the most prominent of which is the glucose sensor. In his review *Electrochemical Glucose Biosensors,* Wang, describes in detail the refinement of the original glucose sensor fabricated by Clark and Lyons in 1962, to today's commercially available disposable electrodes with portable personal readout devices to the potential, continuous non-invasive *in vivo* real time monitoring [83].

Continuing the theme of non-invasive *in vivo* monitoring, Wang also describes the development of a miniaturised flow cell amperometric sensor, which could be inserted into a subjects tear duct. The screen printed microflow detector was first shown to possess the ability to detect catecholamine neurotransmitters norepinephrine and dopamine with the rolled microsensor giving a rapid and sensitive responses to trace levels of the two neurotransmitters (approx. 165 nM norepinephrine and 530 nM dopamine). Further functionalization of the sensor by the addition of a carbon working electrode coated with a film containing glucose oxidase, allowed the sensitive detection (8 μM) of glucose [84].

Wang's group also depict their work with screen printed electrode (SPE) as sensors. Screen printing is a technique that offers a variety of advantages including low-cost mass production, minimal sample volume and low cross-contamination [85]. Au SPEs, coated with a ternary monolayer interface, containing hexanedithiol (HDT), SHCP a thiolated capture probe, and 6-mercapto-1-hexanol (MCH) were shown to offer direct, sensitive detection of nucleic acid hybridization in serum and urine. Modification of the sensing layer further improved the sensors sensitivity with a 10-fold improvement in the signal-to-noise ratio for a 1 nM target DNA compared to common SHCP/MCH binary interfaces. The SPEs also allowed direct quantification of the target DNA down to 25 pM and 100 pM in undiluted serum and urine samples, respectively. These SPEs possessed good anti-fouling properties after extended exposure to raw human serum and urine samples, showing good potential as low cost nucleic acid sensors [86].

In another account Wang *et al.* outline their development of wearable SPEs for the voltammetric detection of trace heavy metal contaminants (copper linear detection from 10 - 90 ppb with a limit of detection (LOD) of 13 ppb) and nitro – aromatic explosives (detection range of 100 - 900 ppb of TNT with a LOD of 42 ppb) in seawater [87]. The modification of SPEs by inkjet printing is further illustrated by a number of other groups including Sriprachuabwong *et al* [88] who describe how the graphene-PEDOT:PSS modified SPEs electrochemical activity to H_2O_2, nicotinamide adenine dinucleotide (NADH) and $K_4Fe(CN)_6$ is significantly enhanced, in comparison to unmodified SPEs.

Microelectrodes are another popular means of miniaturisation as they exhibit increased temporal resolution and current densities, with reduced ohmic drop and charging currents, and high Faradaic to capacitive current ratios. However individually their current output is often overshadowed by electrochemical noise. To counteract this, microelectrode arrays are fabricated, improving sensitivity and facilitating lower detection limits in comparison to macroelectrodes [89]. Thus microelectrodes are one of the most obvious successes in sensor miniaturisation because they are small, able to operate in low sample volumes and with portable instrumentation systems. This has resulted in a significant volume of research into the working mechanism of microelectrodes particularly as they do not experience an ohmic drop during their operation [90-92].

Although the continuous improvement of microfabrication techniques has enabled researchers to fabricate ultramicroelectrodes of increasingly diminutive size [93] it is through the incorporation of novel materials that the greatest advances in sensor development are seen. For example, Dumitrescu *et al.* describe the fabrication of ultramicroelectrodes composed of networks of single-walled CNTs using lithography. The resulting electrodes exhibited low background current and decreased double layer charging in comparison to macroelectrodes, which coupled with their low surface area and capacitance resulted in fast response times in comparison to metal ultramicroelectrodes making them suitable for a variety of applications [94]. Similarly other groups report the application of printed electrodes in environmental and biological analysis [95, 96]

The work of Banks *et al.* highlights the advantages of miniaturised SPEs over conventional carbon electrodes, specifically in terms of cost and time efficiency. SPEs have been shown to possess a greater ease of use, scale of economies and are disposable in nature, requiring reduced volumes of the analyte (μl). This gives them a significant advantage over conventional (solid) carbon electrodes and other fabrication approaches (such as lithography) in sensing applications. The incorporation of nanomaterials to improve electron transfer processes, selectivity and sensitivity will only further enhance the field of electrochemical sensors pushing towards smaller and well-defined geometries in formats such as screen printed microelectrode arrays and 'lab on a chip' approaches with the overall ethos of simpler, cheaper, disposable, scalable and ease of use [85].

It follows that the application of interdigitated electrode arrays (IDAs) can also exploited to further miniaturise sensing systems often through the incorporation of nanocomposites, Liu *et al.* [97] describes the detection of volatile sulphur compounds (associated with a variety of ailments [98-100]) in a subject's breath exhalations. The authors report that the fabricated PANI/AuNPs sensor electrodes exhibited a good response to both H_2S (1.0 ppm) and CH_3SH (1.5 ppm) gases and that in comparison with other sensors, these novel IDA electrodes possess a greater selectivity and sensitivity while also being low cost. IDAs were also the basis of a humidity sensor developed by Cassidy *et al.* [101] where an Ag polymer nanocomposite was shown to give linear detection over a range of 10 – 60 % relative humidity (RH), the resulting sensor was found to be reversible and repeatable with minimal hysteresis.

The design and fabrication of electrochemical sensors remains a vibrant area of research with the miniaturization of electrodes continuing down to the nano regime, allowing the routine measurement of fast electron kinetics at diminutive analyte concentrations. The use of nanomaterials in sensor design continues to increase enabling significant improvements in the analytical performance and utility of biosensors.

4.3. Sensitivity

The sensitivity of a sensor is defined as the slope of the analytical calibration curve for a given analyte. A sensor is sensitive when a small change in analyte concentration causes a large change in the response. Within the linear range of response, the sensitivity is a well-defined value. Improvements in sensitivity of sensors have always been of paramount interest [102]. Immunoassays based on specific antigen-antibody recognition, is a promising approach for selective and sensitive analysis. Electrochemical immunosensors, which combine specific immunoreactions with electrochemical transduction, have attracted growing attention in recent years. They have many advantages including simple instrumentation and operation, high sensitivity and selectivity, and wide linear range. Since enzyme labels provide great signal amplification in the assay and also a large number of antibody–enzyme or antigen–enzyme conjugates are commercially available, the majority of electrochemical immunoassays are based on the use of specific enzyme/substrate couples [103-105]. Sensitivities of ng/ml can routinely be achieved in enzyme-based electrochemical immunosensors. Many interesting signal amplification strategies have been adopted to further improve sensitivity of these sensors [106-108]. For example, an immunosensor with a detection limit of 1.0 pg/mL was developed for Interleukin-6 (IL-6) based on a dual amplification mechanism resulting from Au nanoparticles (AuNP)-poly-dopamine (PDOP) as the sensor platform and multienzyme-antibody functionalized AuNP-PDOP CNT. Nonenzymatic immunosensors have also been reported to be able to achieve very low detection limits. Pichetsumthorn *et al.* [102] reported a nonenzymatic impedimetric biosensor for the small molecule atrazine. They used nanoporous alumina membranes and interfaced them with a printed circuit board. The assay achieved non-faradic detection, i.e. the transfer of charge between the biomolecule and the electrode is mediated without the use of a redox probe. The limit of detection was in the fg/mL regime with a dynamic range of detection spanning from 10 fg/mL to 1 ng/mL. Such high sensitivity was attributed to the confinement of the small molecules in the nanopores of the alumina membranes which resulted in an amplification of the impedance signal measured due to the modulation of the electrical double layer. Another highly sensitive, nonenzymatic electrochemical immunosensor [109] was constructed using a graphene platform combined with mesoporous PtRu alloy as a label for signal amplification. A low detection limit of 9.63 pg/mL using a sandwich-type assay approach for microcystin-LR (MC-LR) was reported. Again, this approach does not rely on enzymes or electrochemical redox mediators to function, but rather on the strong catalytic action of NP-PtRu alloys upon H_2O_2. Anodic stripping voltammetry (ASV) is a highly sensitive method for the analysis of trace concentrations of electroactive species in solution. Detection limits for metal ions at sub-ppb concentrations have been reported. ASV detection of metal nanoparticle labels have been used in both immunosensors [110] and DNA sensors [111]. Au nanoparticles are the most frequently used among all the metal nanoparticle labels available. Recently, Shen & Zhang reported a sensitive immunosensor using ASV for the detection of hepatitis B surface antigen (HBsA) based on copper-enhanced Au nanoparticle labels [112]. A linear range of 0.1–1500 ng/mL with a detection limit of 87 pg/mL was reported which was based on the determination of Cu by ASV.

In the past few years, many efforts have been devoted to improve the sensitivity of metal oxide gas sensors [13]. Sakai *et al.* found that the porous structure of the sensing film played a critical role in the performance of the sensor because it dictated the rate of gas diffusion [113]. Indeed, several groups have found that the particle size of the metal oxide heavily affected the sensitivity of sensor [114, 115]. Although many methods have been reported to synthesize nanoparticles of metal oxides, often they are not stable and may easily aggregate. Besides, nanoparticles may be compactly sintered together during the high temperature film coating process, rendering them nonporous, which is a disadvantage for gas diffusion through them. If porous nanostructures are used as gas sensing materials, the gas sensing properties will be much improved. On the basis of this, nanostructures with many kinds of shapes such as porous nanowires, porous nanotubes and porous nanospheres have been reported to have excellent gas sensitivity properties. A nanocomposite of SnO_2 and multi-walled carbon nanotubes (MWCNT) [116] was exploited to detect persistent organic pollutants (POPs) which possess stable chemical properties and are ordinarily difficult to detect with metal oxides [116]. The ultrasensitive detection of aldrin and dichlorodiphenyltrichloroethane (DDT) was carried out using the nanocomposite of small SnO_2 particles and MWCNTs. The nanocomposite showed a very attractive improved sensitivity compared with a conventional SnO_2 sensor. A LOD of 0.5 ng was achieved for aldrin and DDT, which was attributed to the highly porous 3D structured SnO_2/MWCNT nanocomposites (which enhances gas diffusion).

4.4. Selectivity

Selectivity is defined as the ability of a sensor to detect one specific species even in the presence of a number of other chemical species or interferents. All electrochemical sensors exhibit a high degree of selectivity. This section will discuss recent research to develop highly selective electrochemical sensors and efforts to improve the selectivity of existing sensors through the incorporation enhanced sensing membranes. This section is divided in two parts focusing first on potentiometric and then voltammetric and amperometric sensors.

4.4.1. Potentiometric sensors

As discussed earlier, there are several basic types of potentiometric devices including ISEs, FETs [117, 118]. The ISE is the most representative potentiometric sensor; recently there has been a marked increase in the demand for miniaturised ISEs as interest in the biological application of potentiometric sensors. Recent advances in nanoscale potentiometry were discussed by Bakker and Pretsch with the authors discussing the nature and role of interfacial films on both sides of the ion-selective membrane and summarising the improved general performance of the nano-electrodes in terms of detection limits, biocompatibility, and sensor stability [119]. The application of conducting polymers in potentiometric sensing is discussed widely throughout the literature [120-123] with Faridbod *et al.* [124] reporting the use of conducting polymer-based ISEs to over 60 inorganic ions.

Ultimately the development of new means of ion recognition continues to be the main focus of ISE research, with researchers discussing the synthesis of a range of ionophores targeting silver [125, 126], lead [127] and cobalt [128] among others. In each instance improved selectivity is reported coupled with favourable LODs. Buhlmann's group having reported the success of highly fluorinated liquid phases as sensing materials, detailing the design of numerous ion-selective sensors which incorporate fluorinated membranes [129-131].

Throughout sensor research as a whole, the area with the greatest level of continued consistent development is that which exploits nanomaterials and their novel properties, illustrated by the rapidly growing number of publications in this area. This phenomenon is most evident in the area of electrochemical sensors. Within the field of potentiometric sensors and specifically the design of contacts for all solid state sensors, the potential of nanomaterials such as CNTs and fullerenes has been rigorously examined. Rius *et al.* [132-135] described the development of many sensors based on SW- and/or MW- CNT sensors, including an all-solid-state potassium ion-selective membrane electrode and a perchlorate selective membrane electrode using a MWCNT-derived solid contact.

Recent advancements in the field of potentiometric sensors and especially in terms of ISEs have seen a significant focus on the development and fabrication of sensing membranes particularly in terms of their composition, where the incorporation of CNTs and unique ionophores in both carbon paste electrodes and polyvinylchloride (PVC)-based membranes. Carbon pastes utility as the basis of a sensing membrane is becoming increasingly clear, with the development of a number of sensors incorporating MWCNTs and specific ionophores [136-138].

Trace analysis has always been a strength of electrochemical detection. At present the detection of copper ions has become a priority of industry as many biochemical processes carried out in industrial settings depend on the presence of copper ions. Copper can displace metal ions *in vivo* and the potential of sensing techniques to determine the concentration of copper in a sample are both costly and time-consuming, prompting the fabrication of selective sensors by a number of groups [16, 139, 140]. For example, Petkovic *et al.* [140] discuss the development of a novel Cu^{2+} ISE using a PVC matrix containing N,N′,N″,N‴-tetrakis(2-pyridylmethyl)-1,4,8,11-tetraazacyclotetradecane as the ionophore. A simple construction method affords a cost-effective, rapid measurement of copper in a mixed solution with EDTA. With the additional benefits of a superior working pH range and longer sensor lifetime coupled with a comparable measurement signal in comparison with commercial sensors.

Improved composition of ISE membranes has enhanced the response time for ionic sensing and as detailed above, PVC [140] has proven to be suitable for ISE construction as it allows for ease of construction and improved lifetime/robustness of the sensing matrix. PVC containing ISEs have been utilised in range of industrial studies including medical research and water contamination experiments [137, 141-144]. PVC membranes containing ISEs, have also demonstrated potential as anion selective electrodes. For example, Alvarez-Romero *et al.* [145] discuss the successful development of a graphite-based composite containing epoxy resin, doped with chloride ion doped polypyrrole, as a chloride ion selective electrode. Zahran *et al.* [146] report a novel PVC-based sensor that exhibit selectivity toward halides (chloride and bromine specifically). Similarly Kang *et al* [147] describe the fabrication and characterisation of a PVC based Sc^{3+} octaethylporphyrin composite fluoride selective electrode this is significant as fluoride selective sensors allow for rapid detection of toxic organic fluorophosphates, which can contaminate drinking water.

The sensing of uncharged molecules is a constant challenge for researchers as they are not influenced by ionophores traditionally used to sense anionic or cationic species. The use of molecularly imprinted polymers (MIPs) has been greatly increasing in ISEs [148-151]. For example, Liang *et al.* [148] report the use of a polymeric membrane ISE to detect neutral molecules, utilising a MIP, as the sensing element and an indicator ion that is similar in structure to the targeted neutral species. The sensing protocol incorporated two stages: accumulation of neutral species in the membrane phase followed by the removal of the indicator ion and potential measurement with the ISE. The sensing device was shown to be both selective and sensitive.

The potential of potentiometric sensors is becoming increasingly apparent in biomedical research with potentiometric sensors playing greater roles in pharmaceutical development, disease screening, and disease research. PVC plays a major role in membrane fabrication for ISEs showing great versatility in terms of the variety of species sensed. Kumar *et al.* [152] present a sensor capable of selective determination of domperidone, a clinically useful pharmaceutical for a wide variety of stomach ailments. Typically the pharmaceutical's concentration is measured using a combination of analytical techniques. The authors were able to accomplish this through the use of two separate sensors, made of a PVC membrane and a carbon plate, with the electroactive ion pair domperidonephosphotungstic.

Other groups have also reported the successful fabrication of potentiometric sensors for both drug and biological detection [153-155].

4.4.2. Voltammetric sensors

Voltammetry, or more specifically amperometry is a powerful, potentially highly selective analytical technique [156]. Carbon-based electrodes are widely used in voltammetric studies because of their low cost, availability, stability and processability. This has resulted in the availability of a variety of carbon-based electrodes including GC, carbon paste (CP), polycrystalline boron doped diamond (pBDD), CNTs, and most recently graphene. Carbon-based sensors are easily modified via casting, electrodeposition, functionalisation etc. to impart selectivity. For example coupling nanostructures with high surface area with stripping voltammetry greatly enhances specificity. CNTs are ideal materials for incorporation into electrochemical sensors due to their high surface area, high aspect ratio, and enhanced catalytic properties. This is illustrated by the work of Xu and colleagues who developed a GC electrode modified with a nafion/bismuth/MWCNTs composite film capable of the sensitive detection Pb^{2+} and Cd^{2+} (LODs of 25 and 40 ppt respectively) [157]. The nanocomposite allowed the sensor to exploit bismuth's wide potential window and insensitivity to O_2, as well as the MWCNTs enhanced adsorptive capabilities and high surface area. The authors also noted that there was no interference exhibited by a significant number of anions, with a 500-fold mass ratio of SCN^-, Cl^-, F^-, PO_4^{3-} , SO_4^{2-}, NO_3^- for example showing no influence on the sensing films signal response for Pb(II) and Cd(II).

Again much of sensor research is driven by a focus on a final application, with the demand for improved and continuous environmental monitoring driving a significant portion of sensor development. For example Zima *et al* report the use of numerous CPEs that incorporate nanotube or nanoparticles and are modified both chemically and biologically for the detection of various organic molecules such as pharmaceuticals and environmental pollutants. Within this work the authors describe how the performance of these sensors is enhanced, for example, a Tyrosinase modified CPE detected Cysteine with a LOD of 1 - 8μM with no interference from Ascorbic acid and Uric acid, similarly a CPE modified with Glucose oxidase and gold nanoparticles showed no effects in the presence these common interferents [158].

The incorporation of nanomaterials in the fabrication and development of biosensors has resulted in their improved selectivity. This is particularly evident when considering sensors for the detection of neurotransmitters such as dopamine. Shang and colleagues described the fabrication of a dopamine selective, boron-doped diamond (BDD) electrode [159]. The electrodes sensitivity (LOD of 5 nM) and selectivity was augmented by the anionic nature of the dopant, sulfobutylether-cyclodextrin, which pre-concentrated the dopamine and excluded common anionic interferences like ascorbic acid and uric acid. Yan *et al.* produced a 3,4-dihydroxyphenylacetic acid (DOPAC), a metabolite of dopamine, sensor, through the functionalising of SWCNTs that were adsorbed with cyclophane (a DOPAC recognition element) on to a GC electrode [160]. The sensor demonstrated a significant selectivity to DOPAC as the SWNT transducer coupled with a cyclophane recognition element exhibited a stronger binding affinity for DOPAC over both Ascorbic acid and Dopamine.

The incorporation of bio- materials [161, 162] into electrochemical sensors enables a level of selectivity that is comparable to that seen in nature, with the majority of advancements in biosensors due to the immobilization of a biological material (and hence it's physiochemical properties to an electrode's surface electrode [163, 164]. Mimicking nature selectivity has been attempted with the use of molecularly imprinted polymers but so far has not rivalled the selectivity of biomolecules such as antibodies, nucleic acids, aptamers, etc.

4.5. Multi-analyte detection

Simultaneous detection of multiple analytes is a highly desirable feature of any sensor. Current research is endeavouring to exploit the capability of voltammetric analysis to sensitively, selectively rapidly differentiate between numerous compounds. This is particularly evident in recent developments in biosensors. Recent research highlights the interest in simultaneously quantifying physiologically relevant compounds such as uric acid, ascorbic acid and/or catecholamines [165-167]. A large body of literature describes the production of a variety of sensors with multi-analyte sensing capabilities, exploiting a variety of nanomaterials including CNTs [168-170], individual nanoparticles [171] and polymer nanocomposites [172, 173]

The electronic nose or tongue is an analytical instrument comprising an array of chemical sensors with partial specificity or cross-sensitivity to different components of a mixture/compound, and an appropriate method of pattern recognition and/or multivariate calibration for the data processing. They can quantitatively differentiate between the composition of complex liquids/gases. Such sensors have a number of applications in various industrial areas including the pharmaceutical industry and the food and beverage sector [174], where they can be used to analyse flavour ageing in beverages, quantify bitterness of dissolved compounds, quantify taste masking efficiency of formulations [175] (e.g. tablets, syrups, capsules), analyse medicines stability in terms of taste. Electronic tongue/noses are increasingly studied due to their industrial capabilities coupled with their fast and relatively cheap methodology [176]. For example, Pigani and colleagues detail their development of a poly(3,4 – ethylendioxythiophene) modified electrode that could be incorporated into an electronic tongue. The sensor was utilised in the blind analysis of red wines, (a sample set of 144 red wines of different origin and variety) for classification and calibration purposes. Treating the data obtained from voltammetric measurements using partial least squares analysis allowed the authors to both correlate the calibration procedure results with those from traditional analytical methods and develop classification models for the wines, based on quantitative parameters and qualitative information such as the origin and variety. The sensor may also be used for quality control as it can rapid identify samples that failed to meet threshold limits for SO_2, colour intensity and total polyphenols.

Similar studies by Gutierrez *et al.* [177, 178] in which the authors developed a sensing system for the classification and characterisation of both red and white wines. The sensor consisted of six ISFET sensors, a conductivity sensor, a redox potential sensor, two amperometric electrodes, a gold microelectrode and a microelectrode for sensing electrochemical oxygen demand. The system was shown to be capable of distinguishing between samples according to their grape and even geographical origin the authors also demonstrated the system's ability to differentiate between the mono – varietal samples and their mixtures further emphasise the extent of development sensing has undergone in the last decade and its expanding potential for multiple applications in a variety of industries will only encourage further development of multi-analyte sensing devices.

4.6. Stability

The most desirable sensors are those that retain their characteristics when tested or used under varying conditions and environments, i.e. are those which can function in a myriad of harsh conditions. Where nanomaterials are encompassed in the sensor fabrication, the resulting nanocomposites possess enhanced physical and chemical properties when compared to their constitute components depending on the chemical nature of each component and how they interact. This interaction depends strongly on characteristics of each component, i.e. interface, size, shape and structure. In extreme cases, where there is no or little interaction between the components, the composites' properties should be equivalent to a simple sum of the properties of the individual elements. In cases where the interaction between the constituents is strong, the properties of the composite system can differ substantially from the simple sum of the properties of the individual components. The characteristics of the individual components are lost and new feature arise as a result of the strong interaction [166].

Chik and Xu [179] have described a fabrication method that not only allows the addition of a variety of materials, including metals, semiconductors, and CNTs to an anodised aluminium oxide porous membrane but also, through tuning of the synthesis parameters, controls the nanomaterials morphology. This enabled the authors to engineer the physical properties of the composite by determining the shape, size, composition and doping of the nanostructures, as well as new properties produced by their interaction with the matrix itself. Some of these properties and functions were not intrinsic to the individual nano-elements but were due to the collective behaviour of the nanostructures within the membrane. The novel nanocomposite platform described potentially offers a wide range of applications in various fields including electronics, optics, mechanics, and biotechnologies.

Shi *et al.* [180], summarise a range of chemical sensors that apply the optical principles of nanomaterials for the detection of multiple chemical and biochemical analytes. These include colorimetric biosensors based on gold nanoparticles, gas sensing semi-conductors which exhibit changes to their optical transmittance when exposed to gases such as NO, and CO and optical humidity sensors based on changes in cobalt oxide films' absorbance in the visible wavelength. In this review the authors also report the development of a $SrCO_3$ nanoparticle based ethanol sensor which displayed exceptional durability during a 100 hour reaction with 2000 ppm ethanol at 380 °C.

Chemiresistor sensors have many potential applications, including environmental monitoring [183]. However these sensors can become unstable in extreme environments. The preparation of a nanocomposite can counteract the existing weaknesses of conventional sensors by combining the strengths of nanoparticles with the composite material [184]. This is clearly illustrated in the application of nanocomposites in humidity sensing techniques. Humidity sensing materials can be grouped into two types; ceramics and polymers both possess good chemical and thermal stability, environmental adaptability and a wide range of working temperatures.

Often the sensing mechanisms these materials employ are their surface electrical conductivity or the dielectric constant, which are affected by the adsorption of water vapour. Polymer-based humidity sensing materials possess some advantages in comparison to ceramics; including a higher sensitivity, decreased humidity hysteresis, low cost, flexibility and easy processability [180]. For example, Wang *et al.* [185] describe the improved humidity sensing capability and stability seen by combining nano $BaTiO_3$ with acrylic resin as a nanocomposite humidity sensor. The authors determined that the electrical properties of the sensor, including resistance versus relative humidity, humidity hysteresis, response recovery time and long term stability of the composite sensor were better than that of a sensor composed just of the nano $BaTiO_3$. Similar research has led to multiple reports on chemical polymerisation and electrochemical techniques for the preparation of polymer nanocomposites for gas and humidity sensor applications [186-188].

Similarly the stability of biosensor may be significantly enhanced through the incorporation of nanomaterials [189, 190]. For example, Gopalan *et al.* fabricated an electrochemical biosensor-based on a multi-component composite comprising of MWNT, silica, Nafion and PANI. This blend of these materials within the sensing composite resulted in excellent electrocatalytic activity, sensitivity and stability. Silica gave the composite mechanical stability due to it's rigid 3D porous structure and the inclusion of the MWCNTs was also benefited the sensors mechanical rigidity. Moreover the incorporation of Nafion minimised the brittleness of the pure sol–gel derived silica and enhanced the sensor's long-term stability. The authors utilised this organic-inorganic nanocomposite as a glucose biosensor possessing enhanced stability over time. The sensor displayed a reproducible current response (R.S.D. < 2.2%) for 10 measurements. The stability of the Nafion–silica/MWNT-g-PANI/GOx biosensor was determined by storing the biosensor at 4 °C in PBS (pH 7) for 20 days and monitoring the response current daily for 1.0 mM of glucose at +0.20 V. The biosensor retained its 93% current response after 20 days by loading GOx within the nanocomposite [191].

Similarly Guell and co-workers [192] detail a study of the differing characteristics of three carbon-based electrodes: GC, pBDD, and CNTs. The authors determined that "pristine" CNT networks exhibited background current densities that were 2 orders of magnitude lower than GC and 20 times lower than pBDD for the detection of serotonin, however pBDD electrodes underwent significantly less fouling (minimised by optimisation of the potential range) than the CNTs electrode and exhibited superior stability which is often attributed to the H-surface termination of as - grown pBDD electrodes.

Apetrei *et al.* compared CPEs prepared with graphite, carbon microspheres, or MWCNTs. The electrodes made using carbon microspheres were the most stable (giving consistent results for successive measurement of catechol, with a relative standard deviation of just 3.2%) and gave the best performances in terms of kinetics. Their work also demonstrated that an array combining the three electrode types could discriminate between a number of anti-oxidants depending on their chemical structure and reactivity [193].

Strategies for enhancing the stability of electrochemical sensors continue as advances already made result in the application of sensors in a variety of environments previous considered excessively harsh, e.g. extreme temperatures [194], both high and low pH [195]. Analytical chemists have overcome such challenges by both incorporating biomolecules into sensor platforms and exploiting the desirable properties of novel materials with nanotechnology allowing the development of a variety of disposable and long-life sensors.

5. In summary

Electrochemical sensors have a long and rich history in the field of analytical chemistry contributing to multiple industries. The impact of electrochemical sensors is clearly illustrated by their multiple every day applications and the sheer number of commercially available sensors available. This is evident especially in the field of potentiometric sensors where the further development of ISEs remains the mainstay of recent research. The enhancement of such devices sensing abilities in terms of sensitivity, selectivity, and stability through the incorporation of both novel ionophores into sensor membranes and strategies including analyte preconcentration and background subtraction has allowed the effective detection of cations, anions, and neutral species in challenging environments (conditions with high concentrations of background analytes and biological milieu).

As the demand for sensitive, rapid, and selective determination of analytes continues to grow, so too does the utility of electrochemical sensors. Unlike their spectroscopic and chromatographic counterparts, they are readily adapted for the detection of a wide range of analytes and may be incorporated into robust, portable and/or miniaturised devices while remaining relatively inexpensive.

Despite their ubiquitous presence, electrochemical sensors remain a dynamic field of research especially when coupled with the continued expansion of nanoscience and nanotechnology. Nanomaterials already have extensive applications in electrochemical sensors systems with a significant potential for future development. This is primarily due to the unique and attractive properties that nanoparticles and nanostructures exhibit, the exploitation of which allows the development of electroanalytical systems exhibiting similarly attractive analytical behaviours. A wide variety of metallic and organic nanomaterials have been used to fabricate a diverse range of electrochemical sensing systems, based on their special physical, chemical and even biological properties. Further development of electroanalytical sensor technology, with the discovery and subsequent exploitation of novel properties of nanomaterials will only result in the further evolution of electrochemical sensing platforms.

The majority of modern electrochemical sensor development focuses on the incorporation of both microfabrication and nanofabrication to design sensors of a smaller size and hence lower power demands, of lower cost, and improved portability. The miniaturisation of electrodes continuing down to the nano regime has allowed measurements of fast electron kinetics at very low analyte concentrations. The demand for increasingly low detection levels has been met with the production of arrays. The incorporation of nanostructures has augmented the electrode reactive surface area even with less material used in the physical electrode allowing greater signal to noise ratio for a given analyte.

The development of sensors for the measurement of neurotransmitters is a major trend emerging in voltammetric sensor research, with the sensitivity and selectivity of such sensors being greatly improved due to the development of both measurement techniques and electrode materials. A recent shift in focus from glucose analysis to other physiologically analytes, has driven biosensor development, with DNA hybridization and immunological recognition being the basis of a significant portion of new electrochemical detectors. Again the application of nanomaterials has enabled significant improvements in the analytical performance and utility of biosensors.

In conclusion, the field of electrochemical sensors continues to grow and develop new applications at pace. The incorporation and interaction of unique materials, both nano and biological, remains the focus of a significant portion of research. This will likely continue as the production of sensors with increased specificity and sensors capable of simultaneous determinations with the ability to operate in complex matrixes remains the long term focus of sensor research.

Abbreviations

2-diisopropylaminoethanethiol	DIPAET
Alternating Current	AC
Boron - Doped Diamond	BDD
Carbon Nanotube	CNT
Carbon Paste Electrode	CPE
Coated Wire Electrodes	CWES
Deoxyribonucleic Acid	DNA
Electrochemical Impedance Spectroscopy	EIS
Ethylenediaminetetraacetic acid	EDTA
Femtograms	fg
Glassy Carbon	GC
Glutathione	GSH
Hepatitis B surface Antigen	HBsA
Human Serum Albumin	HSA
Indium Tin Oxide	ITO
Ion Selective Electrode	ISE
Light Addressable Potentiometric Sensors	LAPS
Localised Surface Plasmon Resonance	LSPR
Methyldopa	MDA
Millilitres	ml
Molecularly Imprinted Polymers	MIPs
Nanograms	ng
National Nanotechnology Initiative	NNI
Nicotinamide Adenine Dinucleotide	NADH
Parts per Billion	ppb
Persistent Organic Pollutants	POPs
Poly(3,4-ethylenedioxythiophene)	PEDOT
Poly(styrene-co-acrylic Acid)	PSA
Polyaniline	PANI
Poly-Dopamine	PDOP
Prussian Blue	PB
Quantum Dots	QDs
Room Temperature Ionic Liquids	RTIL
Single Walled Carbon Nanotube	SWCNT
Standard Hydrogen Electrode	SHE

Three Dimensional	3D
3,4-dihydroxyphenylacetic acid	DOPAC
Anodic stripping voltammetry	ASV
Carcinoembryonic Antigen	CEA
Carbon Nanotube Epoxy Composite Electrodes	CNTECE
Chemically Modified Graphene	CMG
Cyclic Voltammetry	CV
Dichlorodiphenyltrichloroethane	DDT
Electrospun Carbon Nanofiber – Modified Carbon Paste Electrode	ECF-CPE
Federation of European Chemical Societies	FECS
Field Effect Transistor	FET
Glucose Oxidase	GOD
Graphene Nanoribbon	GNR
Hexanedithiol	HDT
Immunoglobulin G	IgG
International Union of Pure and Applied Chemistry	IUPAC
Ion Selective Field Effect Transistor	ISFET
Limit of Detection	LOD
Mercapto-1Hexanol	MCH
Microcystin-LR	MC-LR
milli-molar	mM
Multi Walled Carbon Nanotube	MWCNT
Nanoparticles	NPs
Nitrate Reductase	NR
Nuclear Fast Red	NFR
Parts per Million	ppm
Point of Care	PoC
Poly(3,4-ethylenedioxythiophene) : Poly(styrenesulfonic Acid)	PEDOT:PSS
Poly(styrenesulfonic Acid)	PSS
Polycrystalline Boron Doped Diamond	pBDD
Polyvinyl chloride	PVC
Prussian White	PW
Relative Humidity	RH
Screen Printed Electrode	SPE
Specific Thiolated Capture Probe	SHCP
Surface Plasmon Resonance	SPR
Trinitrotoluene	TNT

Author details

Aoife C. Power and Aoife Morrin
National Centre of Sensor Research,
School of Chemical Sciences, Dublin City University, Glasnevin, Dublin, Ireland

6. References

[1] Niinisto L (Chairperson), Working party on Analytical Chemistry (WPAC). In Federation of European Chemical Societies (FECS). 1993. Edingburgh

[2] Cremer M, (1906), Über die Ursache der elektromotorischen Eigenschaften der Gewebe, zugleich ein Beitrag zur Lehre von den polyphasischen Elektrolytketten, Zeitschrift fur Biologie, 47: 562 - 608.

[3] Haber F and Klemensiewicz Z, (1909), Über elektrische Phasengrenzkräfte Zeitschrift für Physikalische Chemie, 67: 385 - 431.

[4] Lubert KH and Kalcher K, (2010), History of Electroanalytical Methods, Electroanalysis, 22: 1937-1946.

[5] Anker JN, *et al.*, (2008), Biosensing with plasmonic nanosensors, Nature Materials, 7: 442-453.

[6] Yi F and La Van DA, (2012), Nanoscale Thermal Analysis for Nanomedicine by Nanocalorimetry, Wiley Interdisciplinary Reviews - Nanomedicine and Nanobiotechnology, 4: 31-41.

[7] Waggoner PS and Craighead HG, (2007), Micro- and Nanomechanical Sensors for Environmental, Chemical and Biological Detection, Lab on a Chip, 7: 1238-1255.

[8] Holford TRJ, Davis F, and Higson SPJ, (2012), Recent trends in antibody based sensors, Biosensors & Bioelectronics, 34: 12-24.

[9] Palchetti I and Mascini M, (2012), Electrochemical Nanomaterial - Based Nucleic Acid Aptasensors, Analytical and Bioanalytical Chemistry, 402: 3103-3114.

[10] Perfezou M, Turner A, and Merkoci A, (2012), Cancer Detection Using Nanoparticle - Based Sensors, Chemical Society Reviews, 41: 2606-2622.

[11] Lei J and Ju H, (2012), Signal Amplification Using Functional Nanomaterials for Biosensing, Chemical Society Reviews, 41: 2122-2134.

[12] Liu Y, Dong X, and Chen P, (2012), Biological and Chemical Sensors Based on Graphene Materials, Chemical Society Reviews, 41: 2283-2307.

[13] Sun YF, *et al.*, (2012), Metal Oxide Nanostructures and Their Gas Sensing Properties: A Review, Sensors, 12: 2610-2631.

[14] Silvester DS, (2011), Recent Advances in the use of Ionic Liquids for Electrochemical Sensing, Analyst, 136: 4871-4882.

[15] Ni YN and Kokot S, (2008), Does Chemometrics Enhance the Performance of Electroanalysis?, Analytica Chimica Acta, 626: 130-146.

[16] Adibi M, Pirali-Hamedani M, and Norouzi P, (2011), Copper Nano-composite Potentiometric Sensor, International Journal of Electrochemical Science, 6: 717-726.

[17] Lin P and Yan F, (2012), Organic Thin-Film Transistors for Chemical and Biological Sensing, Advanced Materials, 24: 34-51.

[18] Scampicchio M, *et al.*, (2012), Electrospun Nonwoven Nanofibrous Membranes for Sensors and Biosensors, Electroanalysis, 24: 719-725.

[19] Kimmel DW, LeBlanc G, Meschievitz ME, and Cliffel D, (2012), Electrochemical Sensors and Biosensors, Analytical Chemistry, 84: 685-707.

[20] Chen W, *et al.*, (2012), Recent advances in electrochemical sensing for hydrogen peroxide: a review, Analyst, 137: 49-58.
[21] Shan X, *et al.*, (2010), Imaging Local Electrochemical Current via Surface Plasmon Resonance, Science, 327: 1363-1366.
[22] (2007), Electrochemical Sensor Analysis, Elsevier:
[23] Zhang XJ, Ju H, and Wang J, (2007), Electrochemical Sensors, Biosensors and their Biomedical Application, Elsevier:
[24] Janata J, (2009), Principles of chemical sensors, Springer Verlag, Dorfrecht, Heidelberg, London, New York:
[25] Bobacka J, Ivaska A, and Lewenstam A, (2008), Potentiometric ion sensors, Chemical Reviews, 108: 329-351.
[26] Bratov A, Abramova N, and Ipatov A, (2010), Recent trends in potentiometric sensor arrays - A review, Analytica Chimica Acta, 678: 149-159.
[27] Faridbod F, Ganjali MR, Dinarvand R, and Norouzi P, (2008), Developments in the field of conducting and non-conducting polymer based potentiometric membrane sensors for ions over the past decade, Sensors, 8: 2331-2412.
[28] Xu L, *et al.*, (2012), An Enantioselective Polyaniline - Coated Membrane Electrode Based on Chiral Salen Mn(III) as Chiral Selector, Analytical Methods, 4: 807-811.
[29] Evtugyn G A, *et al.*, (2008), Selectivity of solid-contact Ag potentiometric sensors based on thiacalix 4 arene derivatives, Talanta, 76: 441-447.
[30] Mattinen U, Rabiej S, Lewenstam A, and Bobacka J, (2011), Impedance Study of the Ion - to - Electron Transduction Process for Carbon Cloth as Solid - Contact Material in Potentiometric Ion Sensors, Electrochimica Acta, 56: 10683-10687.
[31] Rius-Ruiz FX, *et al.*, (2011), Potentiometric Strip Cell Based on Carbon Nanotubes as Transducer Layer: Toward Low - Cost Decentralized Measurements, Analytical Chemistry, 83: 8810-8815.
[32] Pesavento M, *et al.*, (2012), Ion Selective Electrode for Dopamine Based on a Molecularly Imprinted Polymer, Electroanalysis, 24: 813-824.
[33] Li F, *et al.*, (2012), All - Solid - State Potassium - Selective Electrode Using Graphene as the Solid Contact, Analyst, 137: 618-623.
[34] Jia Y, *et al.*, (2011), Bio-Initiated Light Addressable Potentiometric Sensor for Unlabeled Biodetection and its MEDICI Simulation, Analyst, 136: 4533-4538.
[35] Liu Q, *et al.*, (2011), In vitro Assessing the Risk of Drug - Induced Cardiotoxicity by Embryonic Stem Cell - Based Biosensor, Sensors and Actuators B-Chemical, 155: 214-219.
[36] Gupta VK, *et al.*, (2011), Voltammetric techniques for the assay of pharmaceuticals - A review, Analytical Biochemistry, 408: 179-196.
[37] Granado Rico MA, Olivares-Marin M, and Pinilla Gil E, (2009), Modification of carbon screen-printed electrodes by adsorption of chemically synthesized Bi nanoparticles for the voltammetric stripping detection of Zn(II), Cd(II) and Pb(II), Talanta, 80: 631-635.

[38] Fu XC, *et al.*, (2011), Stripping voltammetric detection of mercury(II) based on a surface ion imprinting strategy in electropolymerized microporous poly(2-mercaptobenzothiazole) films modified glassy carbon electrode, Analytica Chimica Acta, 685: 21-28.

[39] Mohadesi A, Teimoori E, Taher MA, and Beitollah H, (2011), Adsorptive Stripping Voltammetric Determination of Cobalt (II) on the Carbon Paste Electrode, International Journal of Electrochemical Science, 6: 301-308.

[40] Somerset V, *et al.*, (2010), Development and Application of a Poly(2,2 '- Dithiodianiline) (PDTDA) - Coated Screen - Printed Carbon Electrode in Inorganic Mercury Determination, Electrochimica Acta, 55: 4240-4246.

[41] Fojta M, Jelen F, Havran L, and Palecek E, (2008), Electrochemical stripping techniques in analysis of nucleic acids and their constituents, Current Analytical Chemistry, 4: 250-262.

[42] Clark LC, (1956), Monitor and Control of Blood and Tissue Oxygen Tensions, Transactions American Society for Artificial Internal Organs, 2: 41-46.

[43] Clark LC and Lyons C, (1962), Electrode Systems for Continuous Monitoring in Cardiovascular Surgery, Annals of the New York Academy of Sciences, 102: 29-45.

[44] Hu CG, *et al.*, (2012), Inkjet Printing of Nanoporous Gold Electrode Arrays on Cellulose Membranes for High-Sensitive Paper-Like Electrochemical Oxygen Sensors Using Ionic Liquid Electrolytes, Analytical Chemistry, 84: 3745-3750.

[45] Morrin A, (2012), Inkjet Printed Electrochemical Sensors, In Korvink JG, *et al.*s Inkjet-based Micromanufacturing, Wiley-VCH: 295-309.

[46] Newman JD and Turner APF, (2005), Home Blood Glucose Biosensors: A Commercial Perspective, Biosensors & Bioelectronics, 20: 2435-2453.

[47] Zhong HA, *et al.*, (2012), Non - Enzymatic Hydrogen Peroxide Amperometric Sensor Based on a Glassy Carbon Electrode Modified with an MWCNT / Polyaniline Composite Film and Platinum Nanoparticles, Microchimica Acta, 176: 389-395.

[48] Guo YJ, Guo SJ, Fang YX, and Dong SJ, (2010), Gold nanoparticle/carbon nanotube hybrids as an enhanced material for sensitive amperometric determination of tryptophan, Electrochimica Acta, 55: 3927-3931.

[49] Ventura DN, *et al.*, (2012), A Flexible Cross - Linked Multi - Walled Carbon Nanotube Paper for Sensing Hydrogen, Carbon, 50: 2672-2674.

[50] Kang E, *et al.*, (2012), Electrospun BMIMPF6/Nylon 6,6 Nanofiber Chemiresistors as Organic Vapour Sensors, Macromolecular Research, 20: 372-378.

[51] Crowley K, *et al.*, (2008), Fabrication of an ammonia gas sensor using inkjet-printed polyaniline nanoparticles, Talanta, 77: 710-717.

[52] Rheaume JM and Pisano AP, (2011), A Review of Recent Progress in Sensing of Gas Concentration by Impedance Change, Ionics, 17: 99-108.

[53] Bataillard P, *et al.*, (1988), Direct Detection of Immunospecies by Capacitance Measurements, Analytical Chemistry, 60: 2374-2379.

[54] Tully E, Higson SP, and Kennedy RO, (2008), The Development of a 'Labeless' Immunosensor for the Detection of Listeria Monocytogenes Cell Surface Protein, Internalin B, Biosensors & Bioelectronics, 23: 906-912.

[55] Caballero D, *et al.*, (2012), Impedimetric immunosensor for human serum albumin detection on a direct aldehyde-functionalized silicon nitride surface, Analytica Chimica Acta, 720: 43-48.

[56] Loo AH, *et al.*, (2012), Impedimetric Immunoglobulin G Immunosensor Based on Chemically Modified Graphenes, Nanoscale, 4: 921-925.

[57] Yang M, *et al.*, (2011), Site - Specific Immobilization of Gold Binding Polypeptide on Gold Nanoparticle - Coated Graphene Sheet for Biosensor Application, Nanoscale, 3: 2950-2956.

[58] Ahmadalinezhad A., Wu G. S., and Chen A. C., (2011), Mediator-free electrochemical biosensor based on buckypaper with enhanced stability and sensitivity for glucose detection, Biosensors & Bioelectronics, 30: 287-293.

[59] Yehezkeli O, Tel-Vered R, Reichlin S, and Willner I, (2011), Nano - Engineered Flavin - Dependent Glucose Dehydrogenase / Gold Nanoparticle - Modified Electrodes for Glucose Sensing and Biofuel Cell Applications, ACS Nano, 5: 2385-2391.

[60] Wipawakarn P, Ju HX, and Wong DKY, (2012), A Label - Free Electrochemical DNA Biosensor Based on a Zr(IV) - Coordinated DNA Duplex Immobilised on a Carbon Nanofibre Chitosan Layer, Analytical and Bioanalytical Chemistry, 402: 2817-2826.

[61] Lahiff E, *et al.*, (2010), The Increasing Importance of Carbon Nanotubes and Nanostructured Conducting Polymers in Biosensors, Analytical and Bioanalytical Chemistry, 398: 1575-1589.

[62] Ansari S. A. and Husain Q., (2012), Potential applications of enzymes immobilized on/in nano materials: A review, Biotechnology Advances, 30: 512-523.

[63] Xiao Y, *et al.*, (2003), "Plugging into Enzymes": Nanowiring of Redox Enzymes by a Gold Nanoparticle, Science, 299: 1877-1881.

[64] Zayats M, Willner B, and Willner I, (2008), Design of Amperometric Biosensors and Biofuel Cells by the Reconstitution of Electrically Contacted Enzyme Electrodes, Electroanalysis, 20: 583-601.

[65] Sia SK and Chin CD, (2011), Analytical Chemistry: Sweet Solution to Sensing, Nature Chemistry, 3: 659-660.

[66] Xiang Y and Lu Y, (2011), Using Personal Glucose Meters and Functional DNA Sensors to Quantify a Variety of Analytical Targets, Nature Chemistry, 3: 697-703.

[67] Lai GS, *et al.*, (2012), Electrochemical Stripping Analysis of Nanogold Label - Induced Silver Deposition for Ultrasensitive Multiplexed Detection of Tumor Markers, Analytica Chimica Acta, 721: 1-6.

[68] Joung CK, *et al.*, (2012), Ultra-Sensitive Detection of Pathogenic Microorganism Using Surface - Engineered Impedimetric Immunosensor, Sensors and Actuators B-Chemical, 161: 824-831.

[69] Cataldo V, Vaze A, and Rusling JF, (2008), Improved detection limit and stability of amperometric carbon nanotube-based immunosensors by crosslinking antibodies with polylysine, Electroanalysis, 20: 115-122.

[70] Lin DJ, *et al.*, (2012), Triple Signal Amplification of Graphene Film, Polybead Carried Gold Nanoparticles as Tracing Tag and Silver Deposition for Ultrasensitive Electrochemical Immunosensing, Analytical Chemistry, 84: 3662-3668.

[71] Peng J, *et al.*, (2012), Calcium Carbonate - Gold Nanocluster Hybrid Spheres: Synthesis and Versatile Application in Immunoassays, Chemistry - a European Journal, 18: 5261-5268.

[72] Zhang DW, *et al.*, (2012), A Label - Free Aptasensor for the Sensitive and Specific Detection of Cocaine Using Supramolecular Aptamer fragments / Target Complex by Electrochemical Impedance Spectroscopy, Talanta, 92: 65-71.

[73] Ohno Y, Maehashi K, and Matsumoto K, (2010), Label - Free Biosensors Based on Aptamer - Modified Graphene Field - Effect Transistors, Journal of the American Chemical Society, 132: 18012-18013.

[74] Renedo OD, Alonso-Lomillo MA, and Martinez MJ, (2007), Recent Developments in the Field of Screen - Printed Electrodes and their Related Applications, Talanta, 73: 202-219.

[75] Gonzalez-Macia L, Smyth MR, and Killard AJ, (2012), A Printed Electrocatalyst for Hydrogen Peroxide Reduction, Electroanalysis, 24: 609-614.

[76] Hu JY, Lin YP, and Liao YC, (2012), Inkjet Printed Prussian Blue Films for Hydrogen Peroxide Detection, Analytical Sciences, 28: 135-140.

[77] Liu H and Crooks RM, (2012), Paper - Based Electrochemical Sensing Platform with Integral Battery and Electrochromic Read-Out, Analytical Chemistry, 84: 2528-2532.

[78] Anastasova-Ivanova S, *et al.*, (2010), Development of miniature all-solid-state potentiometric sensing system, Sensors and Actuators B: Chemical, 146: 199-205.

[79] Mousavi Z, *et al.*, (2011), Comparison of Multi - Walled Carbon Nanotubes and Poly (3,octylthiophene) as Ion - to - Electron Transducers in All - Solid - State Potassium Ion - Selective Electrodes, Electroanalysis, 23: 1352-1358.

[80] Li C, Bai H, and Shi G, (2009), Conducting Polymer Nanomaterials: Electrosynthesis and Applications, Chemical Society Reviews, 38: 2397-2409.

[81] Dahlin AB, *et al.*, (2012), Electrochemical plasmonic sensors, Analytical and Bioanalytical Chemistry, 402: 1773-1784.

[82] Makowski MS and Ivanisevic A, (2011), Molecular Analysis of Blood with Micro- / Nanoscale Field Effect Transistor Biosensors, Small, 7: 1863-1875.

[83] Wang J, (2008), Electrochemical Glucose Biosensors, Chemical Reviews, 108: 814-825.

[84] Kagie A, *et al.*, (2008), Flexible Rolled Thick Film Miniaturized Flow Cell for Minimally Invasive Amperometric Sensing, Electroanalysis, 20: 1610-1614.

[85] Metters JP, Kadara RO, and Banks CE, (2011), New Directions in Screen Printed Electroanalytical Sensors: An Overview of Recent Developments, Analyst, 136: 1067-1076.

[86] Kuralay F, Campuzano S, Haake DA, and Wang J, (2011), Highly Sensitive Disposable Nucleic Acid Biosensors for Direct Bioelectronic Detection in Raw Biological Samples, Talanta, 85: 1330-1337.

[87] Malzahn K, *et al.*, (2011), Wearable Electrochemical Sensors for in situ Analysis in Marine Environments, Analyst, 136: 2912-2917.

[88] Sriprachuabwong C, *et al.*, (2012), Inkjet - Printed Graphene - PEDOT: PSS Modified Screen Printed Carbon Electrode for Biochemical Sensing, Journal of Materials Chemistry, 22: 5478-5485.

[89] Kadara RO, Jenkinson N, and Banks CE, (2009), Screen printed recessed microelectrode arrays, Sensors and Actuators B: Chemical, 142: 342-346.

[90] Amatore C, Oleinick A, and Svir I, (2008), Theoretical Analysis of Microscopic Ohmic Drop Effects on Steady-State and Transient Voltammetry at the Disk Microelectrode: A Quasi-Conformal Mapping Modeling and Simulation, Analytical chemistry, 80: 7947-7956.

[91] Amatore C, Oleinick AI, and Svir I, (2009), Numerical Simulation of Diffusion Processes at Recessed Disk Microelectrode Arrays Using the Quasi-Conformal Mapping Approach, Analytical chemistry, 81: 4397-4405.

[92] Guo J and Lindner E, (2008), Cyclic voltammograms at coplanar and shallow recessed microdisk electrode arrays: Guidelines for design and experiment, Analytical chemistry, 81: 130-138.

[93] Li Y, Bergman D, and Zhang B, (2009), Preparation and Electrochemical Response of 1–3 nm Pt Disk Electrodes, Analytical chemistry, 81: 5496-5502.

[94] Dumitrescu I, Unwin PR, Wilson NR, and Macpherson JV, (2008), Single-Walled Carbon Nanotube Network Ultramicroelectrodes, Analytical chemistry, 80: 3598-3605.

[95] Hallam PM, Kampouris DK, Kadara RO, and Banks CE, (2010), Graphite screen printed electrodes for the electrochemical sensing of chromium (VI), Analyst, 135: 1947-1952.

[96] Khairy M, Kadara RO, Kampouris DK, and Banks CE, (2010), In Situ Bismuth Film Modified Screen Printed Electrodes for the Bio - Monitoring of Cadmium in Oral (Saliva) Fluid, Analytical Methods, 2: 645-649.

[97] Liu C, Hayashi K, and Toko K, (2011), Au Nanoparticles Decorated Polyaniline Nanofiber Sensor for Detecting Volatile Sulfur Compounds in Expired Breath, Sensors and Actuators B: Chemical, 161: 504-509.

[98] Toda K, Li J, and Dasgupta PK, (2006), Measurement of Ammonia in Human Breath with a Liquid - Film Conductivity Sensor, Analytical chemistry, 78: 7284-7291.

[99] Van den Velde S, Nevens F, van Steenberghe D, and Quirynen M, (2008), GC – MS Analysis of Breath Odor Compounds in Liver Patients, Journal of Chromatography B, 875: 344-348.

[100] Hibbard T and Killard AJ, (2011), Breath ammonia analysis: Clinical application and measurement, Critical Reviews in Analytical Chemistry, 41: 21-35.

[101] Power AC, Betts AJ, and Cassidy JF, (2010), Silver Nanoparticle Polymer Composite Based Humidity Sensor, Analyst, 135: 1645-1652.

[102] Pichetsumthorn P, Vattipalli K, and Prasad S, (2012), Nanoporous Impedemetric Biosensor for Detection of Trace Atrazine from Water Samples, Biosensors & Bioelectronics, 32: 155-162.

[103] Neves MMPS, Gonzalez-Garcia MB, Santos-Silva A, and Costa-Garcia A, (2012), Voltammetric Immunosensor for the Diagnosis of Celiac Disease Based on the Quantification of Anti - Gliadin Antibodies, Sensors and Actuators B-Chemical, 163: 253-259.

[104] Rosales-Rivera LC, *et al.*, (2012), Amperometric Immunosensor for the Determination of IgA Deficiency in Human Serum Samples, Biosensors & Bioelectronics, 33: 134-138.

[105] Yu X, Kim SN, Papadimitrakopoulos F, and Rusling JF, (2005), Protein Immunosensor Using Single - Wall Carbon Nanotube Forests with Electrochemical Detection of Enzyme Labels, Molecular Biosystems, 1: 70-78.

[106] Hong CL, *et al.*, (2012), A strategy for signal amplification using an amperometric enzyme immunosensor based on HRP modified platinum nanoparticles, Journal of Electroanalytical Chemistry, 664: 20-25.

[107] Su HL, Yuan R, Chai YQ, and Zhuo Y, (2012), Enzyme-nanoparticle conjugates at oil-water interface for amplification of electrochemical immunosensing, Biosensors & Bioelectronics, 33: 288-292.

[108] Wang GF, *et al.*, (2012), A Supersandwich Multienzyme - DNA Label Based Electrochemical Immunosensor, Chemical Communications, 48: 720-722.

[109] Wei Q, *et al.*, (2011), Nanoporous PtRu Alloy Enhanced Nonenzymatic Immunosensor for Ultrasensitive Detection of Microcystin - LR, Advanced Functional Materials, 21: 4193-4198.

[110] Dequaire M, Degrand C, and Limoges B, (2000), An electrochemical metalloimmunoassay based on a colloidal gold label, Analytical chemistry, 72: 5521-5528.

[111] Authier L, Grossiord C, Brossier P, and Limoges B, (2001), Gold nanoparticle-based quantitative electrochemical detection of amplified human cytomegalovirus DNA using disposable microband electrodes, Analytical Chemistry, 73: 4450-4456.

[112] Shen GY and Zhang Y, (2010), Highly Sensitive Electrochemical Stripping Detection of Hepatitis B Surface Antigen Based on Copper - Enhanced Gold Nanoparticle Tags and Magnetic Nanoparticles, Analytica Chimica Acta, 674: 27-31.

[113] Sakai G, Matsunaga N, Shimanoe K, and Yamazoe N, (2001), Theory of Gas - Diffusion Controlled Sensitivity for Thin Film Semiconductor Gas Sensor, Sensors and Actuators B-Chemical, 80: 125-131.

[114] Xu CN, Tamaki J, Miura N, and Yamazoe N, (1991), Grain-Size Effects on Gas Sensitivity of Porous SNO_2-Based Elements, Sensors and Actuators B-Chemical, 3: 147-155.

[115] Belle CJ, *et al.*, (2011), Size dependent gas sensing properties of spinel iron oxide nanoparticles, Sensors and Actuators B-Chemical, 160: 942-950.

[116] Meng FL, *et al.*, (2010), Nanocomposites of Sub-10 nm SnO_2 Nanoparticles and MWCNTs for Detection of Aldrin and DDT, Analytical Methods, 2: 1710-1714.

[117] Shipway AN, Katz E, and Willner I, (2000), Nanoparticle Arrays on Surfaces for Electronic, Optical and Sensor Applications, ChemPhysChem, 1: 18-52.

[118] Katz E, Willner I, and Wang J, (2004), Electroanalytical and Bioelectroanalytical Systems Based on Metal and Semiconductor Nanoparticles, Electroanalysis, 16: 19-44.

[119] Bakker E and Pretsch E, (2008), Nanoscale Potentiometry, TrAC Trends in Analytical Chemistry, 27: 612-618.

[120] Bobacka J and Ivaska A, (2010), Chemical Sensors Based on Conducting Polymers, Electropolymerization: 173-187.

[121] Long YZ, *et al.*, (2011), Recent Advances in Synthesis, Physical Properties and Applications of Conducting Polymer Nanotubes and Nanofibers, Progress in Polymer Science, 36: 1415-1442.

[122] Lange U and Mirsky VM, (2011), Chemiresistors Based on Conducting Polymers: A Review on Measurement Techniques, Analytica Chimica Acta, 687: 105-113.

[123] Xia L, Wei Z, and Wan M, (2010), Conducting Polymer Nanostructures and their Application in Biosensors, Journal of Colloid and Interface Science, 341: 1-11.

[124] Faridbod F, Norouzi P, Dinarvand R, and Ganjali MR, (2008), Developments in the field of conducting and non-conducting polymer based potentiometric membrane sensors for ions over the past decade, Sensors, 8: 2331-2412.

[125] On JH, *et al.*, (2009), Synthesis of 7- Deoxycholic Amides or Cholanes Containing Distinctive Ion - Recognizing Groups at C3 and C12 and Evaluation for Ion - Selective Ionophores, Tetrahedron, 65: 1415-1423.

[126] Mashhadizadeh MH, Shockravi A, Khoubi Z, and Heidarian D, (2009), Efficient Synthesis of a New Podand and Application as a Suitable Carrier for Silver Ion - Selective Electrode, Electroanalysis, 21: 1041-1047.

[127] Li XG, Ma XL, and Huang MR, (2009), Lead (II) Ion - Selective Electrode Based on Polyaminoanthraquinone Particles with Intrinsic Conductivity, Talanta, 78: 498-505.

[128] Gupta VK, *et al.*, (2008), Electroanalytical studies on cobalt (II) selective potentiometric sensor based on bridge modified calixarene in poly (vinyl chloride), Electrochimica Acta, 53: 5409-5414.

[129] Boswell PG, *et al.*, (2005), Coordinative properties of highly fluorinated solvents with amino and ether groups, Journal of the American Chemical Society, 127: 16976-16984.

[130] Boswell PG, *et al.*, (2008), Fluorophilic Ionophores for Potentiometric pH Determinations with Fluorous Membranes of Exceptional Selectivity, Analytical chemistry, 80: 2084-2090.

[131] Lai CZ, *et al.*, (2009), Fluorous polymeric membranes for ionophore-based ion-selective potentiometry: how inert is Teflon AF?, Journal of the American Chemical Society, 131: 1598-1606.

[132] Crespo GA, Gugsa D, Macho S, and Rius FX, (2009), Solid-contact pH-selective electrode using multi-walled carbon nanotubes, Analytical and Bioanalytical Chemistry, 395: 2371-2376.

[133] Crespo GA, Macho S, and Rius FX, (2008), Ion-selective electrodes using carbon nanotubes as ion-to-electron transducers, Analytical chemistry, 80: 1316-1322.

[134] Crespo GA, Macho S, Bobacka J, and Rius FX, (2008), Transduction mechanism of carbon nanotubes in solid-contact ion-selective electrodes, Analytical chemistry, 81: 676-681.

[135] Parra EJ, *et al.*, (2009), Ion - Selective Electrodes Using Multi - Walled Carbon Nanotubes as Ion - to - Electron Transducers for the Detection of Perchlorate, Analyst, 134: 1905-1910.

[136] Faridbod F, *et al.*, (2010), Ho^{3+} carbon paste sensor based on multi-walled carbon nanotubes: Applied for determination of holmium content in biological and environmental samples, Materials Science and Engineering: C, 30: 555-560.

[137] Ganjali MR, *et al.*, (2010), Determination of Pb^{2+} ions by a modified carbon paste electrode based on multi-walled carbon nanotubes (MWCNTs) and nanosilica, Journal of hazardous materials, 173: 415-419.

[138] Norouzi P, *et al.*, (2010), ER $^{3+}$ Carbon Paste Electrode Based on New Nano-Composite, International Journal of Electrochemical Science, 5: 367-376.

[139] Mashhadizadeh MH, Khani H, and Shockravi A, (2010), Used a New Aza - Thia - Macrocycle as a Suitable Carrier in Potentiometric Sensor of Copper (II), Journal of Inclusion Phenomena and Macrocyclic Chemistry, 68: 219-227.

[140] Petković BB, *et al.*, (2010), A Copper (II) Ion-Selective Potentiometric Sensor Based on N, N′, N ″, N‴ Tetrakis (2,pyridylmethyl) 1, 4, 8, 11, Tetraazacyclotetradecane in PVC Matrix, Electroanalysis, 22: 1894-1900.

[141] Abbaspour A, Mirahmadi E, Khalafi-Nejad A, and Babamohammadi S, (2010), A highly selective and sensitive disposable carbon composite PVC-based membrane for determination of lead ion in environmental samples, Journal of hazardous materials, 174: 656-661.

[142] Hosseini M, Abkenar SD, Ganjali MR, and Faridbod F, (2011), Determination of zinc (II) ions in waste water samples by a novel zinc sensor based on a new synthesized Schiff's base, Materials Science and Engineering: C, 31: 428-433.

[143] Motlagh MG, Taher MA, and Ahmadi A, (2010), PVC membrane and coated graphite potentiometric sensors based on 1-phenyl-3-pyridin-2-yl-thiourea for selective determination of iron (III), Electrochimica Acta, 55: 6724-6730.

[144] Zamani HA, *et al.*, (2011), Quantitative Monitoring of Terbium Ion by a Tb^{3+} Selective Electrode Based on a New Schiff's Base, Materials Science and Engineering: C, 31: 409-413.

[145] Álvarez-Romero GA, *et al.*, (2010), Development of a Chloride Ion-Selective Solid State Sensor Based on Doped Polypyrrole-Graphite-Epoxy Composite, Electroanalysis, 22: 1650-1654.

[146] Zahran EM, *et al.*, (2009), Triazolophanes: A New Class of Halide - Selective Ionophores for Potentiometric Sensors, Analytical Chemistry, 82: 368-375.

[147] Kang Y, *et al.*, (2010), Development of a Fluoride-Selective Electrode based on Scandium (III) Octaethylporphyrin in a Plasticized Polymeric Membrane, Bulletin of the Korean Chemical Society, 31: 1601-1608.

[148] Liang RN, Song DA, Zhang RM, and Qin W, (2010), Potentiometric Sensing of Neutral Species Based on a Uniform Sized Molecularly Imprinted Polymer as a Receptor, Angewandte Chemie, 122: 2610-2613.

[149] Madunić D, Sak-Bosnar M, and Matešić-Puač R, (2011), A New Anionic Surfactant - Sensitive Potentiometric Sensor with a Highly Lipophilic Electroactive Material, International Journal of Electrochemical Science, 6: 240-253.

[150] Washe AP, Macho S, Crespo GA, and Rius FX, (2010), Potentiometric Online Detection of Aromatic Hydrocarbons in Aqueous Phase Using Carbon Nanotube - Based Sensors, Analytical Chemistry, 82: 8106-8112.

[151] Zhuiykov S, Kats E, and Marney D, (2010), Potentiometric Sensor Using Sub - Micron Cu_2O - Doped RuO_2 Sensing Electrode with Improved Antifouling Resistance, Talanta, 82: 502-507.

[152] Girish Kumar K, Augustine P, and John S, (2010), Novel potentiometric sensors for the selective determination of domperidone, Journal of applied electrochemistry, 40: 65-71.

[153] Abounassif A, Al-Hadiya BM, and Mostafa GAE, (2010), PVC Matrix Membrane Sensors for Potentiometric Determination of Arecoline, Instrumentation Science and Technology, 38: 165-177.

[154] Alizadeh T and Akhoundian M, (2010), A novel potentiometric sensor for promethazine based on a molecularly imprinted polymer (MIP): The role of MIP structure on the sensor performance, Electrochimica Acta, 55: 3477-3485.

[155] Vlascici D, *et al.*, (2010), Manganese (III) Porphyrin - Based Potentiometric Sensors for Diclofenac Assay in Pharmaceutical Preparations, Sensors, 10: 8850-8864.

[156] Killard AJ and Smyth MR (2006), Electrochemical Immunosensors, In Grimes CA, *et al.*s Encyclopedia of Sensors, American Scientific Publishers, Pennsylvania:

[157] Xu H, *et al.*, (2008), Ultrasensitive Voltammetric Detection of Trace Lead (II) and Cadmium (II) Using MWCNTs Nafion/Bismuth Composite Electrodes, Electroanalysis, 20: 2655-2662.

[158] Zima J, Švancara I, Barek J, and Vytřas K, (2009), Recent Advances in Electroanalysis of Organic Compounds at Carbon Paste Electrodes, Critical Reviews in Analytical Chemistry, 39: 204-227.

[159] Shang F, *et al.*, (2009), Selective Nanomolar Detection of Dopamine Using a Boron - Doped Diamond Electrode Modified with an Electropolymerized Sulfobutylether - β - Cyclodextrin - Doped Poly (N - Acetyltyramine) and Polypyrrole Composite Film, Analytical chemistry, 81: 4089-4098.

[160] Yan J, *et al.*, (2008), An Electrochemical Sensor for 3, 4 - Dihydroxyphenylacetic Acid with Carbon Nanotubes as Electronic Transducer and Synthetic Cyclophane as Recognition Element, Chemical Communications: 4330-4332.

[161] Siangproh W, Dungchai W, Rattanarat P, and Chailapakul O, (2011), Nanoparticle - Based Electrochemical Detection in Conventional / Miniaturized Systems and their Bioanalytical Applications: A Review, Analytica Chimica Acta, 690: 10-25.

[162] Guo S and Wang E, (2011), Noble metal nanomaterials: Controllable synthesis and application in fuel cells and analytical sensors, Nano Today, 6: 240-264.

[163] Ispas CR, Crivat G, and Andreescu S, (2012), Review: Recent Developments in Enzyme-Based Biosensors for Biomedical Analysis, Analytical Letters, 45: 168-186.

[164] Moyo M, Okonkwo JO, and Agyei NM, (2012), Recent Advances in Polymeric Materials Used as Electron Mediators and Immobilizing Matrices in Developing Enzyme Electrodes, Sensors, 12: 923-953.

[165] Atta NF, El-Kady MF, and Galal A, (2010), Simultaneous determination of catecholamines, uric acid and ascorbic acid at physiological levels using poly (N-methylpyrrole)/Pd-nanoclusters sensor, Analytical Biochemistry, 400: 78-88.

[166] Gholivand MB and Amiri M, (2009), Preparation of Polypyrrole/Nuclear Fast Red Films on Gold Electrode and Its Application on the Electrocatalytic Determination of Methyl-dopa and Ascorbic Acid, Electroanalysis, 21: 2461-2467.

[167] Zachek MK, *et al.*, (2009), Simultaneous Decoupled Detection of Dopamine and Oxygen Using Pyrolyzed Carbon Microarrays and Fast - Scan Cyclic Voltammetry, Analytical Chemistry, 81: 6258-6265.

[168] Noroozifar M, Khorasani-Motlagh M, and Taheri A, (2010), Preparation of Silver Hexacyanoferrate Nanoparticles and its Application for the Simultaneous Determination of Ascorbic Acid, Dopamine and Uric Acid, Talanta, 80: 1657-1664.

[169] Rastakhiz N, Kariminik A, Soltani-Nejad V, and Roodsaz S, (2010), Simultaneous Determination of Phenylhydrazine, Hydrazine and Sulfite Using a Modified Carbon Nanotube Paste Electrode, International Journal of Electrochemical Science, 5: 1203-1212.

[170] Ensafi AA and Karimi-Maleh H, (2010), Modified multiwall carbon nanotubes paste electrode as a sensor for simultaneous determination of 6-thioguanine and folic acid using ferrocenedicarboxylic acid as a mediator, Journal of Electroanalytical Chemistry, 640: 75-83.

[171] Ghorbani-Bidkorbeh F, Shahrokhian S, Mohammadi A, and Dinarvand R, (2010), Simultaneous voltammetric determination of tramadol and acetaminophen using carbon nanoparticles modified glassy carbon electrode, Electrochimica Acta, 55: 2752-2759.

[172] Kalimuthu P and John SA, (2010), Simultaneous Determination of Ascorbic Acid, Dopamine, Uric Acid and Xanthine Using a Nanostructured Polymer Film Modified Electrode, Talanta, 80: 1686-1691.

[173] Ulubay S and Dursun Z, (2010), Cu Nanoparticles Incorporated Polypyrrole Modified GCE for Sensitive Simultaneous Determination of Dopamine and Uric Acid, Talanta, 80: 1461-1466.

[174] Ghasemi-Varnamkhasti M, *et al.*, (2011), Electronic and bioelectronic tongues, two promising analytical tools for quality evaluation of non alcoholic beer, Trends in Food Science & Technology, 22: 245-248.

[175] Woertz K, Tissen C, Kleinebudde P, and Breitkreutz J, (2010), Rational Development of Taste Masked Oral Liquids Guided by an Electronic Tongue, International Journal of Pharmaceutics, 400: 114-123.

[176] Riul Jr A, Dantas CAR, Miyazaki CM, and Oliveira Jr ON, (2010), Recent Advances in Electronic Tongues, Analyst, 135: 2481-2495.

[177] Gutiérrez M, *et al.*, (2011), Application of an E-Tongue to the Analysis of Monovarietal and Blends of White Wines, Sensors, 11: 4840-4857.

[178] Gutiérrez M, *et al.*, (2010), Hybrid electronic tongue based on optical and electrochemical microsensors for quality control of wine, Analyst, 135: 1718-1725.

[179] Chik H and Xu JM, (2004), Nanometric superlattices: non-lithographic fabrication, materials, and prospects, Materials Science and Engineering: R: Reports, 43: 103-138.

[180] Shi J, *et al.*, (2004), Recent Developments in Nanomaterial Optical Sensors, TrAC Trends in Analytical Chemistry, 23: 351-360.

[181] Khanna VK, (2008), Nanoparticle - Based Sensors, Science, 58: 608-616.

[182] Ellis DI and Goodacre R, (2006), Metabolic fingerprinting in disease diagnosis: biomedical applications of infrared and Raman spectroscopy, Analyst, 131: 875-885.

[183] Naydenova I, Jallapuram R, Toal V, and Martin S, (2008), A Visual Indication of Environmental Humidity Using a Color Changing Hologram Recorded in a Self - Developing Photopolymer, Applied Physics Letters, 92: 031109.

[184] Ahir SV, Huang YY, and Terentjev EM, (2008), Polymers with aligned carbon nanotubes: Active composite materials, Polymer, 49: 3841-3854.

[185] Wang J, Lin Q, Zhou R, and Xu B, (2002), Humidity Sensors Based on Composite Material of Nano - $BaTiO_3$ and Polymer RMX, Sensors and Actuators B: Chemical, 81: 248-253.

[186] Novak BM, (1993), Hybrid Nanocomposite Materials - Between Inorganic Glasses and Organic Polymers, Advanced Materials, 5: 422-433.

[187] Selampinar F, *et al.*, (1995), A Conducting Composite of Polypyrrole II. As a Gas Sensor, Synthetic metals, 68: 109-116.

[188] Patil D, Patil P, Seo YK, and Hwang YK, (2010), Poly (o - Anisidine) Tin Oxide Nanocomposite: Synthesis, Characterization and Application to Humidity Sensing, Sensors and Actuators B: Chemical, 148: 41-48.

[189] Barbadillo M, *et al.*, (2009), Gold nanoparticles-induced enhancement of the analytical response of an electrochemical biosensor based on an organic–inorganic hybrid composite material, Talanta, 80: 797-802.

[190] Mao S, *et al.*, (2009), Ultrafast Hydrogen Sensing Through Hybrids of Semiconducting Single - Walled Carbon Nanotubes and Tin Oxide Nanocrystals, Physical Chemistry Chemical Physics (PCCP), 11: 7105-7110.

[191] Gopalan AI, *et al.*, (2009), An electrochemical glucose biosensor exploiting a polyaniline grafted multiwalled carbon nanotube/perfluorosulfonate ionomer–silica nanocomposite, Biomaterials, 30: 5999-6005.

[192] Güell AG, Meadows KE, Unwin PR, and Macpherson JV, (2010), Trace voltammetric detection of serotonin at carbon electrodes: comparison of glassy carbon, boron doped diamond and carbon nanotube network electrodes, Physical Chemistry Chemical Physics (PCCP), 12: 10108-10114.

[193] Apetrei C, Apetrei IM, Saja JAD, and Rodriguez-Mendez ML, (2011), Carbon paste electrodes made from different carbonaceous materials: application in the study of antioxidants, Sensors, 11: 1328-1344.

[194] Radecka M., *et al.*, Nanocrystalline TiO 2/SnO 2 composites for gas sensors, Journal of Thermal Analysis and Calorimetry: 1-6.

[195] Sung T.W. and Lo Y.L., (2012), Highly sensitive and selective sensor based on silica-coated CdSe/ZnS nanoparticles for Cu< sup> 2+</sup> ion detection, Sensors and Actuators B: Chemical:

Chapter 7

Shape Classification for Micro and Nanostructures by Image Analysis

F. Robert-Inacio, G. Delafosse and L. Patrone

Additional information is available at the end of the chapter

1. Introduction

In many fields considering information at microscale or nanoscale requires to achieve an automatic shape classification by image analysis, because of the amount of particles, on the one hand, and because of their size and reachability. This shape classification can be set up in 2D or in 3D. In 2D, images are generally provided by a scanning electron microscope (SEM). In 3D, different means can be used, such as reconstruction from a set of 2D sections or as direct tridimensionnal acquisition, for example by an atomic force microscope (AFM). Depending on the microscope resolution, shape studies can be lead either at nanoscale or at microscale. In this chapter, several shape parameters are defined and examples are given in two different fields of application: nanoelectronics and nuclear power. Both applications are achieved in 2D and 3D.

2. Shape classification: state of the art

In pattern recognition, an important field of interest concerns shape classification. Most of the methods used to reach this aim, consist in evaluating some geometrical features of the shapes under study, in order to determine which kind of shape family they belong to. For example, symmetry according to a point [23][24][47][48] can be measured by several means, such as Minkowski's, Besicovitch's, Winternitz's or Blaschke's coefficients [5][19]. Other applications require the evaluation of the symmetry degree according to an axis [31][42][43]. These studies are realized in 2D, 3D or arbitrary dimensions. And finally, circularity, elongation and so on [21][26], can be estimated in order to classify shapes. Classification methods can be shared in two different sets: the first one processing on natural images in gray levels or in color, in 2D or 3D, and the second one on binary images describing well-defined objects. In the last case, natural images need to be pre-processed in order to identify shapes to study. Our method is one of this kind, and we are going to study the similarity between 2D or 3D objects.

3. Shape parameters in 2D and 3D

Several means can be used to describe shapes: circularity parameter, fractal degree of a 2D shape, polar study of the 2D boundary, peak dispersion on a surface in 3D or study of height variations in 3D.

3.1. Circularity parameters

Let us start these studies by considering a classical shape parameter enabling to estimate the circularity degree of an object in two dimensions. It is based on a simple observation about circular objects. For a disk D of radius R, the perimeter $P(D)$ is given by:

$$P(D) = 2\pi.R \tag{1}$$

and the surface area $S(D)$:

$$S(D) = \pi.R^2 \tag{2}$$

It yields that the circularity parameter CP, defined, for any object X, by:

$$CP(X) = \frac{P(X)^2}{4\pi.S(X)} \tag{3}$$

is equal to 1, if and only if X is a disk.

Shape	Perimeter P	Surface area S	CP
Square X_1	$4.l$	l^2	$\frac{4}{\pi} \approx 1.273$
Hexagon X_2	$6.l$	$3.l^2.cos\frac{\pi}{6}$	$\frac{3}{\pi.cos\frac{\pi}{6}} \approx 1.103$
Hexagon $X_2(\alpha)$ with a triangular concavity	$(6+\alpha).l$	$\left(3-\frac{\alpha^2}{2}\right).l^2.cos\frac{\pi}{6}$	$\frac{(6+\alpha)^2}{2\pi(6-\alpha^2).cos\frac{\pi}{6}}$
Rectangle X_3	$2.l.(1+t)$	$t.l^2$	$\frac{(1+t)^2}{\pi t} > \frac{4}{\pi}$
Rectangle $X_3(\alpha)$ with a square concavity	$2l.(1+t)$	$l^2.(t-\alpha^2)$	$\frac{(1+t)^2}{\pi(t-\alpha^2)}$
Rectangle $X'_3(\alpha)$ with a rectangular concavity	$P(X_3)+\alpha$	$S(X_3)-\frac{\alpha(k-\alpha)}{4}$	$\frac{[P(X_3)+\alpha]^2}{4\pi.S(X_3)-\pi.\alpha(k-\alpha)}$

Table 1. CP values for the set of shapes of Fig. 7

In this way, we can see that CP increases when X becomes less close to a disk in shape. Fig. 8 illustrates the results given in Table 1. The shapes under study are:
• A square X_1 of edge length l,
• A regular hexagon X_2 of edge length l,
• A regular hexagon $X_2(\alpha)$ with a triangular concavity (Fig. 1)
• A rectangle X_3 of height length l, and width length $t.l$, $t > 1$
• A rectangle $X_3(\alpha)$ with a square concavity (Fig. 2)
• A rectangle $X'_3(\alpha)$ with a rectangular concavity (Fig. 3)

Let us notice that the following relation is true, whatever the connected shape X:

$$CP(X) \geq CP(ch(X)) \tag{4}$$

where $ch(X)$ denotes the convex hull of X.

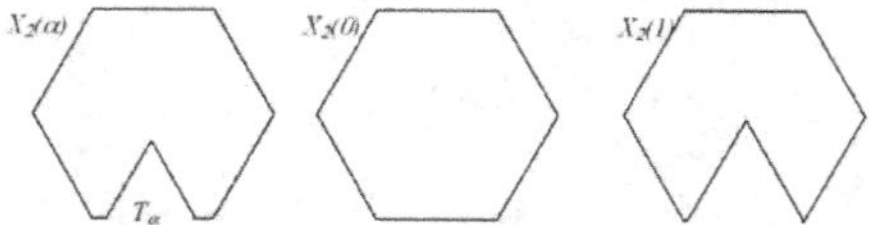

Figure 1. Shape family of non convex sets $X_2(\alpha)$ of convex hull defined by a regular hexagon, with a single triangular concavity T_α (equilateral triangle of edge length $\alpha.l$ with $0 \leq \alpha \leq 1$). $X_2(0)$ and $X_2(1)$ are the two extreme shapes of the family

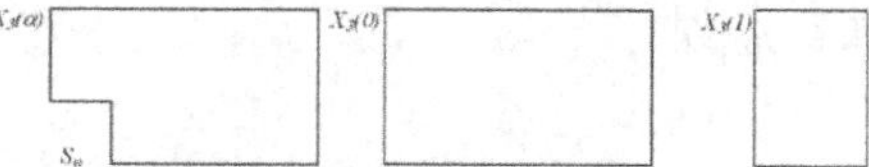

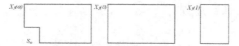

Figure 2. Shape family of non convex sets $X_3(\alpha)$ of convex hull defined by a rectangle, with a single square concavity S_α (square of edge length $\alpha.l$ with $0 \leq \alpha \leq 1$). $X_3(0)$ and $X_3(1)$ are the two extreme shapes of the family

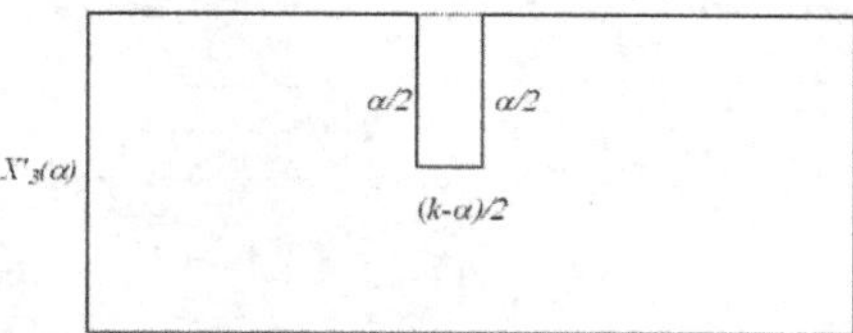

Figure 3. Definition of X_α according to a scalar value α, $\alpha \in [0, k]$

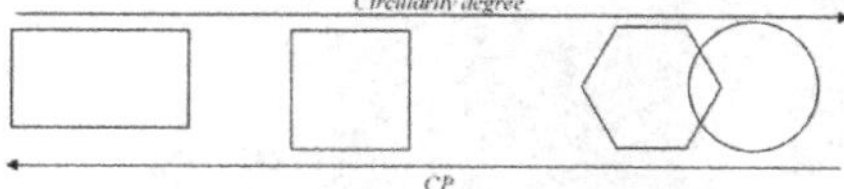

Figure 4. Classification of shapes X_1, X_2 and X_3 from the less to the most circular one, according to CP

Actually, $P(X) \geq P(ch(X))$ and $S(X) \leq S(ch(X))$. So we get the relation of eq. 4.

Furthermore, the parameter CP presents some interesting features such as invariance under affine transformations such as translations, scaling and rotations. It is then quite obvious that shape studies can be led without taking into account neither shape position and orientation, nor scale. As the parameter value is only related with the perimeter and the surface area, it can be interesting to pay more attention to the evolution of CP, when shapes are damaged by one or more concavities. We call concavity of X a connected set of points included in $ch(X)$ but not in X, having a non-empty intersection between its boundary and those of $ch(X)$. In this way, holes are not allowed. Furthermore, we assume that two concavities of a same set X have an empty intersection.

The concavity number is not really preponderant as it is always possible to find an equivalent concavity to a set of alterations, in terms of perimeter and surface area. In other words, the estimation of CP is the same, when considering a single concavity C or a set of concavities C_i, as far as the two following assumptions are satisfied:

$$P(C) = \sum_{i=1}^{N} P(C_i) \tag{5}$$

and

$$P\left(Bd(ch(X)) \cap Bd(C)\right) = P\left(Bd(ch(X)) \cap \left(\bigcup_{i=1}^{N} C_i\right)\right) \tag{6}$$

where $Bd(X)$ is the boundary of X, $S(C) = \sum_{i=1}^{N} S(C_i)$.

In this way, if X is a convex shape:

$$CP(X \backslash C) = CP\left(X \backslash \bigcup_{i=1}^{N} C_i\right) \tag{7}$$

A second circularity parameter can be defined by considering the ratio between radii of the

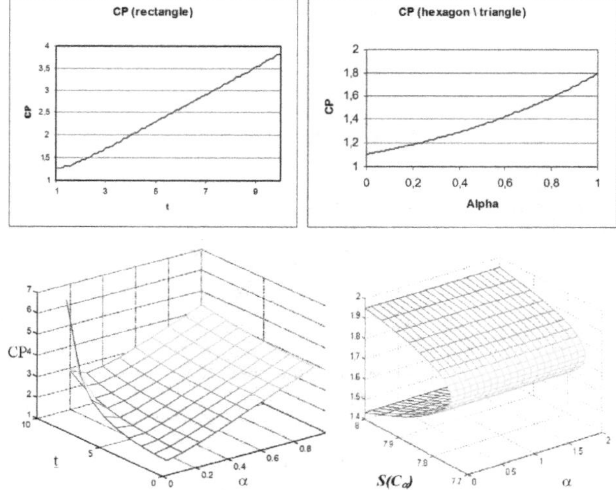

Figure 5. a) Evolution of CP values according to the elongation t of rectangular shapes, b) Evolution of CP according to α for a regular hexagon with a triangular concavity, c) Evolution of CP according to α and t for a rectangle with a square concavity, d) Evolution of CP according to α and $S(C_\alpha)$ for a rectangle with a square concavity

inscribed and circumscribed disks to the shape under study. If $r(X)$ and $R(X)$ are respectively the inscribed and circumscribed disks radii, then $CP2(X)$ is defined as follows:

$$CP2(X) = \frac{r(X)}{R(X)} \tag{8}$$

Fig. 6 illustrates the inscribed disk and circumscribed disk positions for a shape. While the circumscribed disk is unique, the inscribed disk can be located at different positions.

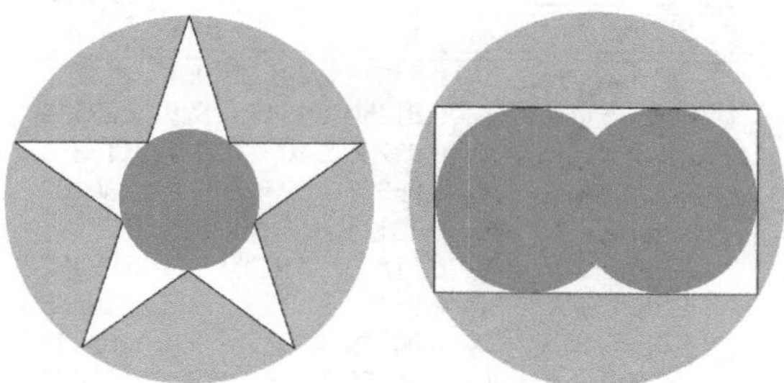

Figure 6. Disks associated to a shape (star or rectangle): in blue, circumscribed disk, in red, inscribed disk

$CP2$ values belong to [0,1] as they are positive values and $CP2(X)$ is equal to 1 if X is a disk. In this case, the circumscribed and the inscribed disks are equal to X, that gives the maximal value for $CP2$.
The inscribed disk can be easily defined by using a euclidean distance map [16] that gives for each point of X its distance to the boundary. The maximal values of such a map are located at inscribed disk centers. The circumscribed disk can be determined by using the circumscribed disk algorithm (see section 3.2.2).

3.2. Shape classification according to a given shape

In order to achieve shape classification according to a reference shape, a similarity parameter P is defined for any pair of convex shapes (X, Y) by considering two scale ratios. The first one gives the smallest homothetic set of X containing Y, and the second one, the smallest homothetic set of Y containing X. Thus, let us define the following function $S_X(Y)$:

$$\begin{aligned} S_X : \mathbb{K} &\longrightarrow \mathbb{R}^+ \\ Y &\mapsto S_X(Y) \end{aligned} \tag{9}$$

where

$$S_X(Y) = inf\{k > 0; Y \subset_t k.X\} \tag{10}$$

$\subset_t$ means "included in, regardless to any translation". And then, a definition of the similarity parameter can be:

$$P(X,Y) = \frac{S_Y(X)}{S_X(Y)} \cdot \frac{\mu(X)}{\mu(Y)} \tag{11}$$

where μ is the surface area measure. The parameter properties are the following:

1. if $X \subset_t Y$ then $P(X,Y)$ belongs to $]0,1]$
2. P is invariant by translation
3. P is invariant by scaling
4. if X and Y are of the same shape regardless to a positive scale ratio then $P(X,Y) = P(Y,X) = 1$
5. $P(X,Y) = P(Y,X)^{-1}$

Shape X	X_i	$S_X(Y)$	$S_Y(X)$	$\mu(X)$	$P_Y(X)$
Square	X_1	0.84375	2.09677	961	0.18736
Pentagon	X_2	0.81250	1.45455	829	0.22436
Parallelogram	X_3	0.93750	2.40741	891	0.16811
Rectangle 1	X_4	0.96875	1.69767	1118	0.30909
Rectangle 2	X_5	0.78125	2.56000	825	0.12198
Triangle	X_6	0.71875	2.41026	585	0.08452
Ellipse 1	X_7	0.68750	2.17500	691	0.10585
Ellipse 2	X_8	0.84375	4.33333	365	0.03445
Circle	X_9	0.96875	1.58537	1320	0.39087
Regular hexagon	X_{10}	0.93750	1.39130	1320	0.43094

Table 2. Results of computation for $P_Y(X)$, $\mu(Y) = 2064$

In order to set up a process of shape classification, the two most interesting properties of the similarity parameter P are given in (1) and (4). In other words, under the assumption that $X \subset_t Y$, we can compute a similarity parameter P_Y, estimating for each convex set X, its similarity degree to Y. By this way, the values reached by P_Y belong to the interval $]0,1]$. So we are able to arrange in order the sets X under study, from the less similar to Y to the most similar. Let us study the following example. The reference shape Y that has to be compared with the others is a hexagon. Then, let us consider ten arbitrary shapes X_i, such as a square, a pentagon, a parallelogram, two rectangles, a triangle, two ellipses, a circle and a regular hexagon. We compute the parameter P_Y for each of these ten arbitrary shapes, and, by this way, we are able to define a partial order for this set of shapes, according to their similarity degree to the hexagon Y. Fig. 7a describes the convex sets Y and X_i and Table 2 gives the computational results. In this way, the previous results can be illustrated as shown in Fig. 7b.

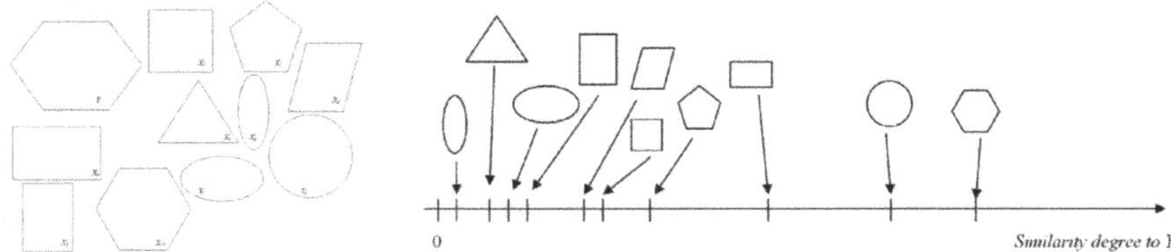

(a) Convex sets under study: hexagon Y and arbitrary shapes X_i

(b) Convex sets under study: hexagon Y and arbitrary shapes X_i

Figure 7. Shape study on a set of simple sets

About this classification, we can note that the closest set in shape to the hexagon Y is the regular hexagon. Furthermore, the second one is the circle, which is close to every regular polygon. And finally, the third one is the rectangle oriented in the horizontal direction, like the hexagon Y. On the opposite side, the less close shape to Y is an ellipse oriented in the vertical direction. Thus, the similarity parameter seems to take into account some geometrical features such as the orientation and the elongation. Then, if the reference shape is a disk P estimates the circularity degree. Furthermore this parameter can be used in 2D or in 3D. It is enough to change the area measure μ into a volume measure to obtain the tridimensionnal version of the parameter P.

3.2.1. *Implementation of the similarity parameter*

In order to implement the similarity parameter, we must set up fast and easy algorithms determining the features of circumscribed convex shapes. In this way, the scale ratios required for the similarity degree estimation will be easily evaluated. Assuming that we have at our disposal an algorithm called $CircumRatio(X, Y)$, determining the scale ratio to apply to a convex set X to circumscribe it to a convex set Y, the algorithm estimating the similarity degree between two shapes is the following:

Algorithm

$k = CircumRatio(X, Y)$
$k' = CircumRatio(Y, X)$
Compute the two surface areas of X and Y
Compute the similarity degree of X and Y

First of all, let us consider the circumscribed disk algorithm that allows the user to compute and design the minimal disk containing a given planar object X. The extension of this algorithm to convex sets is the algorithm used to evaluate the scale ratios, and then, the similarity parameter.

3.2.2. *The circumscribed disk algorithm*

The main results concerning the circumscribed disk algorithm are:

Proposition 1

X is a compact set of $\mathbb{R}^2$ and $B(x, R]$, the closed ball whose center is x and radius R. The following assertions are equivalent:

$$X \subset B(x, R] \tag{12}$$

$$x \in \bigcap_{y \in X} B(y, R] \tag{13}$$

$$x \in \bigcap_{y \in \partial X} B(y, R] \tag{14}$$

$$x \in \bigcap_{y \in E(X)} B(y, R] \tag{15}$$

where $E(X)$ is the extreme point set of the convex hull of X. Let us recall that a point a of a convex body K is extreme if $\forall x, y \in K, a = \frac{x+y}{2} \Rightarrow x = y$. Fig. 8 is an illustration of the four previous assertions.

Then, the features of the circumscribed disk to any compact set are given by the following formulae:

$$R(X) = inf\{R > 0, B(O, R] \ominus \partial X \neq \emptyset\} \tag{16}$$

$$a_X = \{a(X)\} = B(O, R(X)] \ominus \partial X \tag{17}$$

$R(X)$ is the radius of the circumscribed disk and $a(X)$ its center.

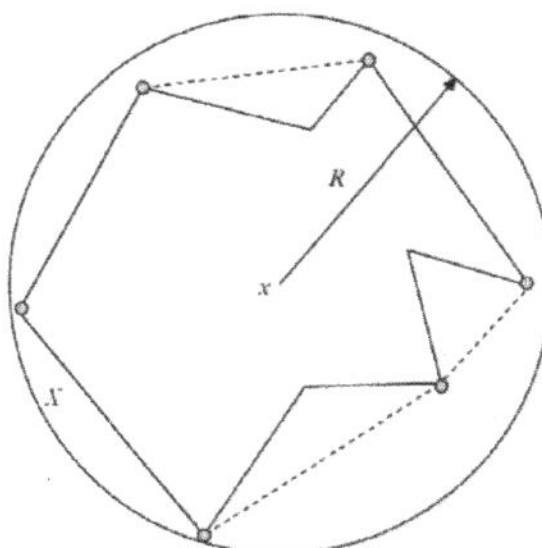

Figure 8. Illustration of Proposition 1: in plain line, the compact set X, in dashed line, its convex hull $C(X)$ and in gray, its extreme points $E(X)$

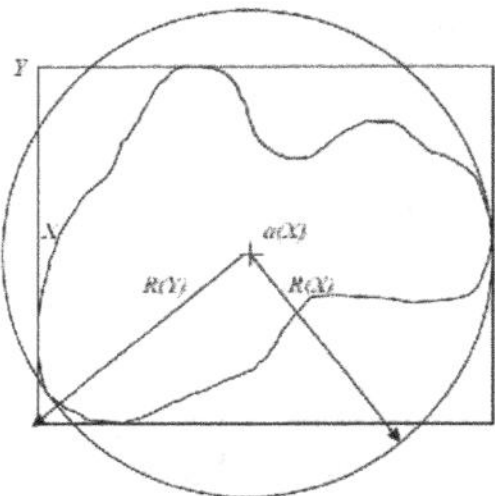

Figure 9. $R(X) \leq R(Y)$

Furthermore, let $R \geq R(X)$, $a(X)$ is the ultimate eroded set of $B(O, R] \ominus \partial X$ by a closed ball of elementary radius $\epsilon > 0$. Let us recall that the ultimate eroded set of a set X by a set Y, is the last non-empty eroded set of X in succesive erosions by Y. In other words, there exists a positive integer n such that $X \ominus nY \neq \emptyset$ and $X \ominus (n+1)Y = \emptyset$. In this way, $X \ominus nY \neq \emptyset$ is the ultimate eroded set of X by Y. From the previous properties, an algorithm can be described by the following steps:

Algorithm 1

1) Acquire a compact set X
2) Search $C(X)$, convex hull of X
3) Search $E(X)$, extreme point set of $C(X)$
4) For each point x of $E(X)$
4-1) Draw the circle whose center is x and radius R
4-2) Fill this circle with respect to the intersection set found yet
4-3) Delete all points which do not belong to the new intersection
5) Ultimate eroded set of the disk intersection

Some classical algorithms exist for the convex hull computation [2][4]. For the determination of extreme points of this convex hull, we can consider that the convex hull is a polygon when working in a discrete space. Then, it is sufficient to determine for each vertex of this polygon if it is extreme or not, in other words, if it is convex or not.

The computation of the ultimate eroded set enables not only to find the location of the center but also to evaluate the radius value. In fact, if $B(O, R_0]$ is the closed ball used to obtain the ultimate eroded set of $B(O, R] \ominus \partial X$, then the radius $R(X)$ is given by:

$$R(X) = R - R_0 \tag{18}$$

The initial value R can be chosen as half the length of the diagonal line of a circumscribed rectangle Y to the given compact X. This value is actually the radius value $R(Y)$ of the circumscribed disk to the rectangle Y. As the rectangle Y contains the compact set X, X is also included in its circumscribed disk and $R(Y)$ is greater or equal to $R(X)$ (Fig. 9).

We can remark that solving the difficult problem of computing the circumscribed disk to any object is reduced to obtain the inscribed disk in a convex set. Actually, the convex set is $B(O, R] \ominus \partial X$ (that is convex because it is the intersection of convex sets) and, when computing the ultimate eroded set by $B(O, R_0]$, the resulting point is the center of the inscribed disk into $B(O, R] \ominus \partial X$ (the radius of this disk is R_0).

3.2.3. *Extension to the third dimension*

The similarity parameter can easily be extended to the third dimension by replacing the surface area measure by the volume measure of 3D objects. The definition is the following:

$$P(A, B) = \frac{S_A(B)}{S_B(A)} \cdot \frac{V(A)}{V(B)} \tag{19}$$

where V is the volume measure. The circumscribed disk algorithm is based on theoretical results that are also true in 3D. That is why the process can be set up in 3D.

3.3. Circularity parameter and fractal degree

The definition of CP and the previous considerations on some particular shapes leads us to the conclusion that the circularity parameter CP is able to estimate a ratio giving indications about the fractal degree [44] on shapes in 2D. It is indeed based on the computation of a ratio between the square of the perimeter and the surface area of the shape under study. In this case, CP increases according to the perimeter while the surface area remains constant. This feature is appropriate to deduce information about the fractal degree for the shape under study, as fractal objects have the feature that their perimeters tend to an infinite value, whereas their surface areas are bounded. So the greater this parameter value, the more fractal the shape, in other words, shapes of a same surface area are more and more fractal while their perimeter increases. Fig. 10 shows that CP increases in an exponential way, when the perimeter value grows. The initial value is computed for a disk of radius of 5. Then the perimeter is approximately equal to 31.4159 and the surface area to 78.5398, and obviously, CP is equal to 1.

As CP depends on the square of the perimeter, its values tend to $+\infty$ when the perimeter increases.

$$CP(X) = f\left(P(X)^2\right) \Rightarrow lim_{P(X) \to +\infty} CP(X) = +\infty \tag{20}$$

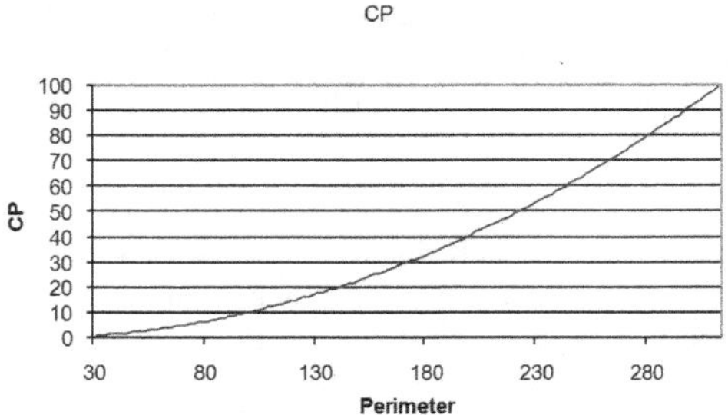

Figure 10. Evolution of CP according to the perimeter

That induces that CP estimates the fractal degree of a shape X. In other words, the higher the CP value, the more fractal the shape X.

3.4. Polar study of the boundary of 2D objects and estimation of the circularity degree

This section presents a local estimation of the circularity degree for objects in 2D [37]. The method consists in drawing the boundary of the shape under study, in polar coordinates, according to an appropriate centre. This centre is chosen as the best-centred point amongst the set of centres of the inscribed disks in the shape. In other words, this centre is the centre of an inscribed disk and it is the closest one to the centroid of the shape under study. We will call this particular point the polar centre of the shape. In the three following examples, this polar centre coincides with the centroid.

The polar graphs corresponding to a square, a rectangle and a regular hexagon (Fig. 11) show a visual estimation of the circularity degree of those three shapes, whereas the standard deviation measures the dispersion of radius values around the average radius. The higher the standard deviation, the less circular the shape. Table 3 gives numerical results of average and standard deviation computation for the three shapes under study. Obviously, if the shape is a disk, the average value is equal to its radius and the standard deviation to 0.

Shape	Square	Rectangle	Regular hexagon
Average radius value	5.7381	4.2000	4.5611
Standard deviation of radius values	0.6311	1.0596	0.2031
Maximal distance to the average value	1.3330	1.7000	0.4389

Table 3. Visual estimation of the circularity degree from polar graphs

>From Table 3, shapes can be ordered from the less circular to the most circular, according to the standard deviation values: rectangle, square and regular hexagon. This result coincides with those obtained from the circularity parameter CP.

Fig. 11 shows that the circularity degree of a shape is closely linked to the amplitude of its polar graph. In other words, the maximal distance measured to the average value of each graph is another way to estimate the circularity degree, that is equivalent to the standard deviation evaluation, as shown in Table 3.

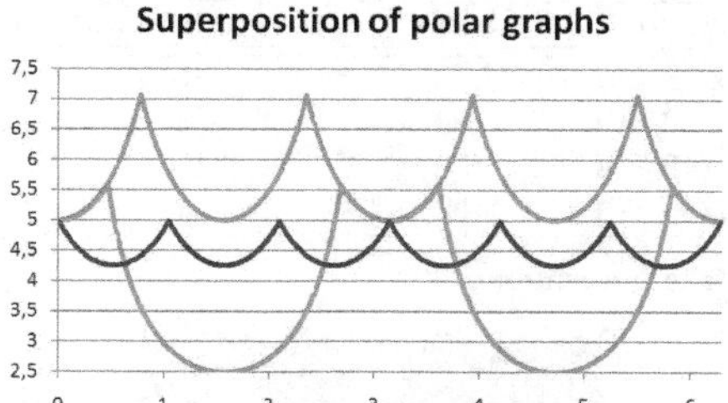

Figure 11. Boundary in polar coordinates of the three previous shapes (coarse angle sampling step). In dark blue, the hexagon, in light blue, the square and in pink, the rectangle

3.5. Study of peak dispersion on a surface in 3D

A surface can present several particular spots located at local maxima in terms of height. These spots are called peaks and they are characterised by their summit (single point) that is higher than its neighbours. Detection of peaks is achieved by considering a square sliding window of 5 pixels width. In order to study the peak dispersion on a given surface, the first step of the process consists of the computation of an Euclidean distance map [16], giving at each point of the surface its distance to the boundary. This distance map is computed on a projection of the surface in the plane (O, x, y), defined in the introduction section as the plane of the corresponding grey-level image. Fig. 12 shows such a bounded surface whose projection is a square of 80 pixel-width and Fig. 13 is the associated Euclidean distance map. On this map, every point belonging to the surface outside is set to 0 and points of the surface are set to their Euclidean distance from the boundary. Then the surface is scanned in order to detect peaks as local maxima. Such a task is achieved by considering decreasing height values. Then each point is checked to determine if it is a local maximum or not. Three main peaks can be detected on this surface. Their features as well as those of the 2D centroid are given in Table 4. The centroid is taken under consideration as it is a point representative of the whole shape in 2D. Furthermore, in this case, it is also the point where the distance map reaches its maximal value. In other words, this point can be considered as the centre of an inscribed disk into the 2D shape.

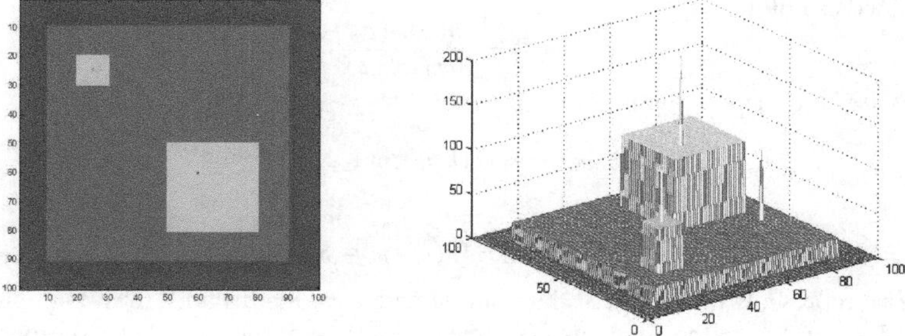

Figure 12. Overview of a bounded surface (left) and its 3D representation (right)

Point	Location (X, Y)	Height	Distance to the boundary	Peak parameter
Peak 1	(25, 25)	150	16	0.2927
Peak 2	(60, 60)	200	31	0.7561
Peak 3	(40, 80)	100	11	0.1341
Centroid in 2D	(50, 50)	100	41	0.5000

Table 4. Features of main peaks and centroid

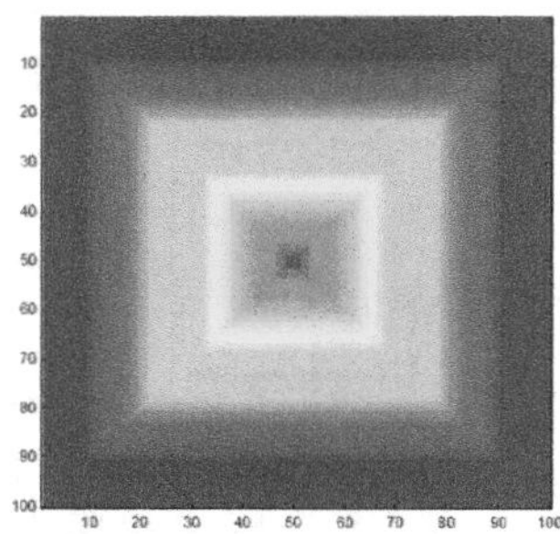

Figure 13. Associated Euclidean distance map

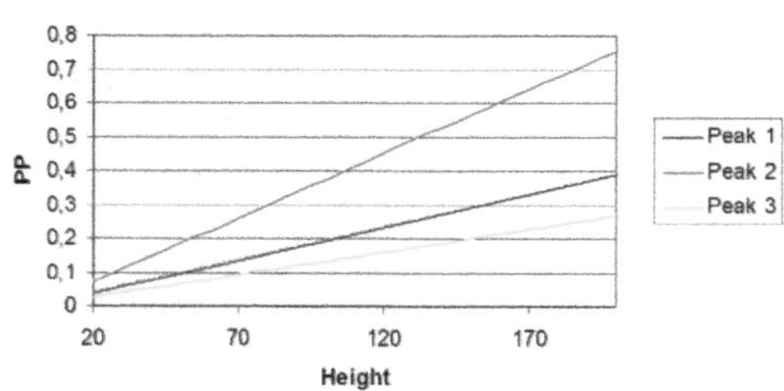

Figure 14. Influence of peak height on PP values

>From the previous features, a parameter can be associated to each peak. This peak parameter PP is defined as follows:

$$PP(p) = \frac{h(p).d(p)}{hmax.dmax} \tag{21}$$

for a peak p of the shape X, with:

$$hmax = max\{h(m), m \in X\} \tag{22}$$

and

$$dmax = max\{d(m), m \in X\} \tag{23}$$

where $h(m)$ represents the height at the point m and $d(m)$ the distance from the point m to the boundary of X. In the previous case, $dmax = 41$ and $hmax = 200$. The parameter PP is maximal (equal to 1) if the highest peak coincides with the point of maximal value on the Euclidean distance map. This parameter enables to estimate a combination between height and centring. Its values vary between 0 and 1 and a value close to 0 represents a non-significant peak in terms of height or centring, or both. The peak dispersion is then established by arranging in order all the significant peaks according to the value of parameter PP. In order to characterize the shape X, the parameter $\overline{PP}$ is determined as the average value of PP on the peaks p of X:

$$\overline{PP}(X) = \frac{1}{N} \sum_{p \in X} PP(p) \tag{24}$$

where N is the number of peaks p of X. In the previous case:

$$\overline{PP}(X) = 0.3943 \tag{25}$$

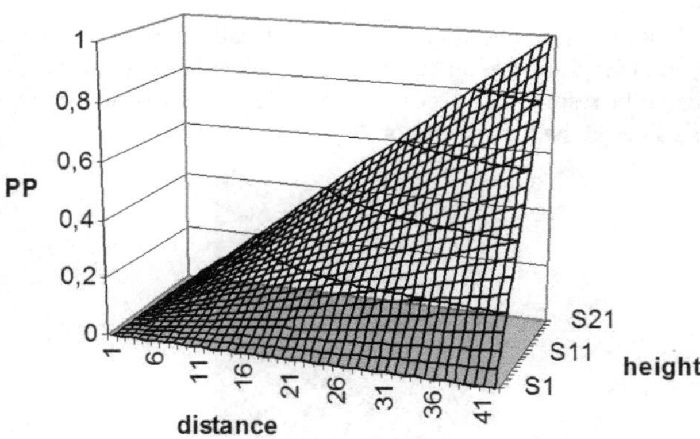

Figure 15. Influence of peak height and distance to the boundary on PP values

The estimation of PP at the maximal point of the distance map gives an evaluation of the centring degree of the shape in terms of height. Fig. 14 shows how the peak height influences the value of PP. The distance to the boundary is then preponderant to determine the coefficient of the slope. On Fig. 15, the influence of the distance and the height of a peak has been drawn for a maximal distance value $dmax$ of 41 and a maximal peak height $hmax$ of 200 ($S1$ corresponds to a height of 0, $S11$ to 100 and $S21$ to 200), that are the characteristic values of the bounded surface X. This graph is representative of the surface and gives the features of centring and height of an arbitrary peak, whatever its position or height. It is then enough to read the corresponding value on the graph to determine PP.

Furthermore, a coefficient of centring CC can be estimated at each peak p of X, described by the following formula:

$$CC(p) = \frac{d(p)}{dmax} \tag{26}$$

Table 5 gives the CC values for the main peaks of X.

Peak (X, Y)	(25, 25)	(60, 60)	(40, 80)
CC	0.3902	0.7561	0.2683

Table 5. Coefficients of centring for the three peaks of X

3.6. Study of height variations

The simplest value to be computed from a bounded surface X in order to estimate height variation is the average value $\bar{h}(X)$.

$$\bar{h}(X) = \frac{1}{card(X)} \sum_{m \in X} h(m) \tag{27}$$

where $card(X)$ is the number of points belonging to X. In this case:

$$\bar{h}(X) = 32.6795 \tag{28}$$

The second way to study height variations consists in locally combining height and distance to the boundary, in order to link height and centring. That has been done in Fig. 16. Each point of this image is the result of the product of the height by the distance to the boundary of X. Table 6 shows values at the main peaks of X.

Figure 16. Local estimation of height variation and centring

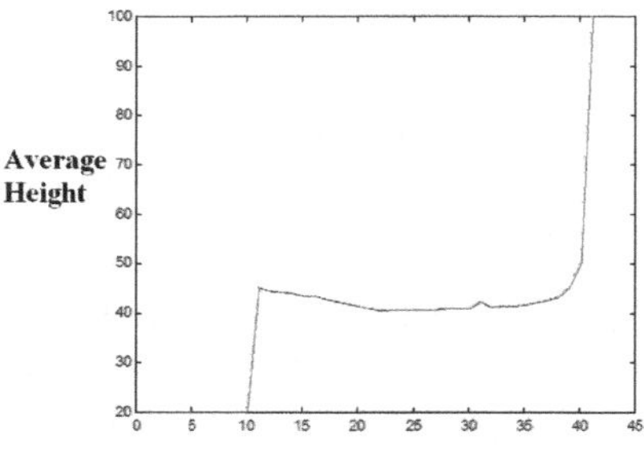

Figure 17. Graph of height variations (VH) according to the distance from the boundary of X

This time too, the peak located at (60, 60) is the most significant. But what is interesting on Fig. 16 is the spot of yellow points where values are around 4000. They are representative of

Peak (X,Y)	(25, 25)	(60, 60)	(40,80)
Local estimation of height and centring combination	2400	6200	1100

Table 6. Height and centring combination (HCC) at main peaks of X

a region of well-centred points at a medium height. The HCC values are more significant in this region than they are at peaks located at $(25,25)$ and $(40,80)$. For example, the centroid located at $(50,50)$ has a HCC value equal to 4100. But we must remember that the centroid is also the maximal point for the Euclidean distance map in this particular case.

Finally, the variation in height VH according to the distance to the edges can be estimated by computing the average value on a given level of the distance map. In other words, we consider all points at a given distance to the boundary and we compute the average height. This can be done for each distance value from 1 to the maximal value. Fig. 17 is the graph of $VH(X)$.

Unfortunately, such a graph (Fig. 17) makes the assumption that the bounded surface under study has an isotropic behaviour, as it is based on an average value computation that does not take into account orientation. It should be better to draw an equivalent surface in polar coordinates to unfold the original surface. The resulting information would show the variations of height according to the angle, with respect to the polar centre.

The peak dispersion on the considered bounded surface is then determined by setting up a process computing the heights (from the original grey-level image) and locations (from the Euclidean distance map) of the main peaks. That will define a simplified map of the surface, including only main information in terms of height.

4. Application 1: sphericity of microscopic particles

4.1. Context

V/HTR (Very/High Temperature Reactors) are advanced nuclear power reactors that employ spherical particles made of an uranium kernel surrounded by four different layers (Fig. 18) [14, 27, 33]. During fuel particle manufacturing, the coating process may generate non spherical particles and/or ceramic layers with abnormal thickness. For such particles, the fuel performances are highly decreased [12, 18]. That is why, it is necessary to set up a characterizing tool allowing, on the one hand, to estimate the thickness and sphericity degree of these fuel particles at each step of the fabrication and, on the other hand, to measure a statistically representative sample taken from fabrication batches containing millions of particles [13]. A classification of fuel particles has to be established in order to reject through statistical control HTR particles batches presenting too serious faults.

4.2. Aim of the study

HTR nuclear fuel particles are made of a kernel of uranium oxide of 500 ţm covered by a first layer of porous pyrocarbon (fission gas tank) and a layer of silicium carbide sandwiched between two layers of dense pyrocarbon. The aim of this study is to measure some geometrical features of such particles, in particular, the kernel diameter and sphericity, and the thickness

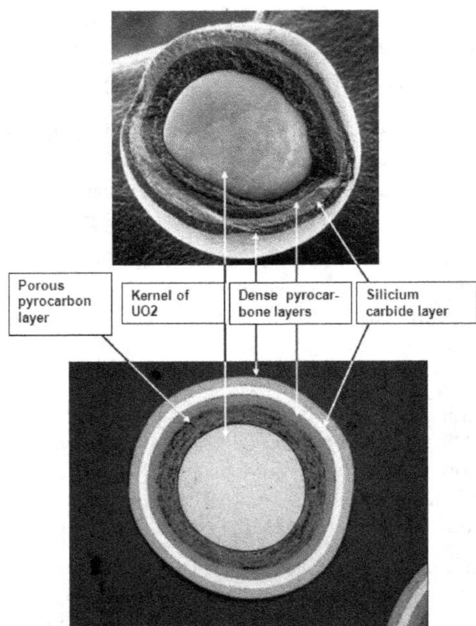

Figure 18. Fuel particle in 3D and polished cross-section

of each layer. The presented method is based on different thresholding processes enabling to extract the kernel or one of the layers from the image [37]. Then, an Euclidean distance map is computed on the binary images, giving for the one corresponding to the kernel its diameter, and for the others, an estimation of the layer thickness. An original shape parameter evaluating the sphericity degree of the kernel is then deduced from the distance map and its value is compared to the shape parameter computed from Féret's diameters. Furthermore, some information related to thickness is extracted from the layer distance maps. For example, the maximum, minimum and average thicknesses are computed from these maps, but a regular set of thickness values can be obtained according to angular sampling. This method enables an automatic control of the particle features in order to classify them in two sets : satisfactory particles and rejected particles [37].

The Euclidian distance mapping [16] is a method that estimates the minimal distance between each pixel of the background and the nearest object. It works on binary images (Fig. 19). The first distance map is computed inside the kernel in order to obtain for each of its points, its distance to the outside, and then, the centre of inscribed disk of the kernel. This centre is determined as the maximal value of the distance map, which represents the radius of the inscribed disk, and it is also considered as the real centre of the particle. After that, a second distance map is computed in order to calculate the kernel radius, by temporarily considering the centre of the particle as the only object of the image. Thus, this new distance map gives a set of concentric circles centred at the kernel centre (Fig. 20). The use of a border

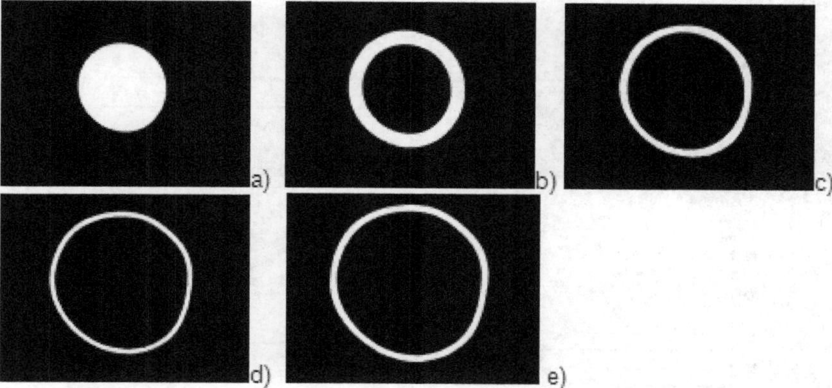

Figure 19. Binary images corresponding to a) the kernel, b) porous pyrocarbon, c) inner pyrocarbon, d) silicium carbide and e) outer pyrocarbon

follow-up method will reveal the kernel local radius by looking at the intersection between the binary image of the kernel and the last distance map. Finally, the process is extended to the surrounding layers by computing four other distance maps on the different binary images representing the different layers. The half thickness of a layer is then determined as the locally maximal values on the distance map. In this way, we obtain thickness measurements in every direction, that allow us to draw the polar graphs.

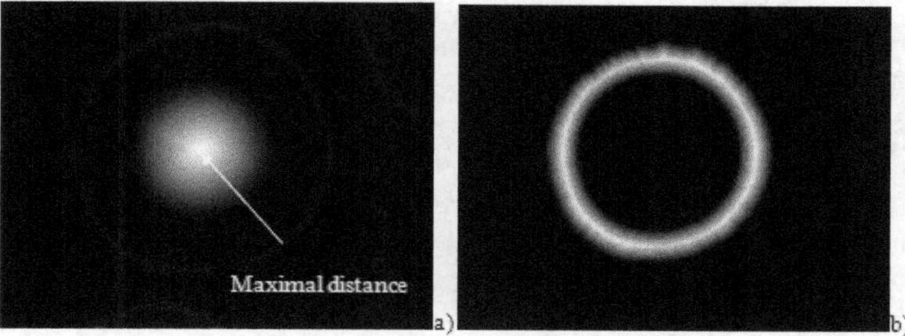

Figure 20. Distance maps associated with a) the kernel and b) the SiC layer

4.3. Experimental results

Fig. 21 shows polar graphs of the different layers and Fig. 22 presents the links between the particle and the polar graphs. Futhermore, Table 7 gives thickness results at the equatorial cross-section, in other words, theoretical and computed thickness results of each element of the particle. The minimum and maximum values for each element are useful in order to detect major local deformation of the particle, while the average value is more representative of the particle regularity.

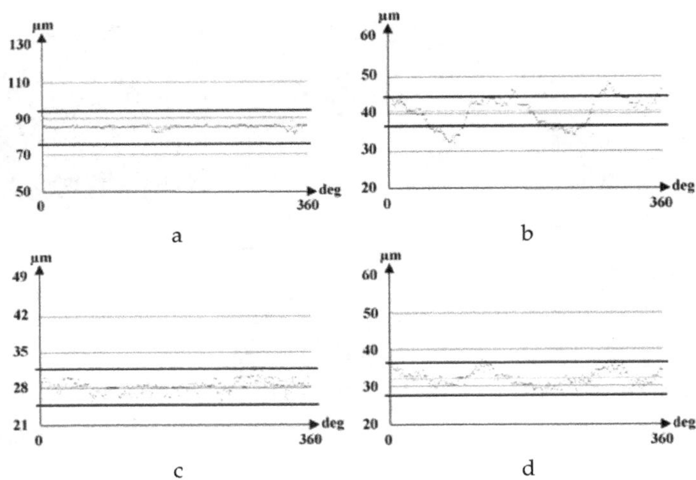

Figure 21. Real deposit of the different layers : a) Porous PyC, b) inner PyC, c) SiC and d) outer PyC

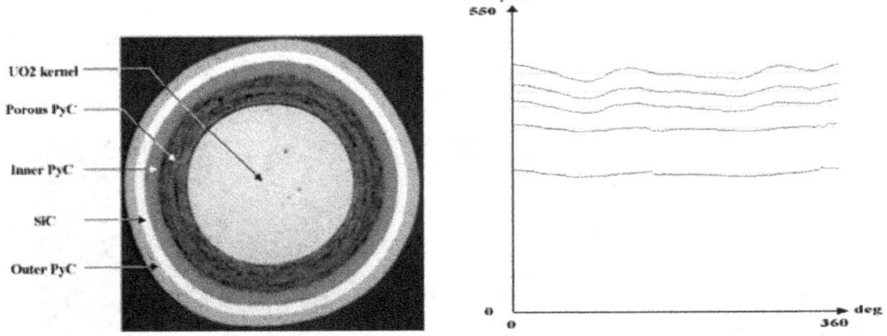

(a) Polished cross-section of a HTR fuel particle

(b) Polar graph representing the kernel and the four layers boundaries of the particle

Figure 22. HTR fuel particle study

Element	Thickness at the Equatorial Cross-section (μm)			
	Maximum	Minimum	Average	Theory
Kernel (radius)	263.38	247.96	253.18	250±20
Porous PyC	88.03	81.14	85.29	90±20
Inner PyC	48.47	32.29	40.61	40±10
SiC	31.44	20.21	28.40	35±7
Outer PyC	37.20	27.22	32.12	40±10

Table 7. Thickness results appearing at the equatorial cross-section

Table 8 gives the computed spherical results of a particle at different states of reconstruction. The results are given in percentages where 100 % means that the particle is perfectly circular.

Element	Sphericity degree (%)	
	CP	*CP2*
Kernel (K)	96.02	94.81
K + Porous PyC	97.24	96.14
K + Inner PyC	94.57	94.00
K + SiC	94.27	94.25
K + Outer PyC	94.06	93.80

Table 8. Sphericity degree results at different levels of the particle

5. Application 2: shape analysis on molecular islands of carbon chains in nanoelectronics

5.1. Self-assembly

Self-assembly is an universal phenomenon, responsible for the structural organization of a system without external intervention. The molecular self-assembly is studied in order to obtain nanometer-sized structures in a bottom-up approach. This one consists in building a structure from individual base elements step by step.

Such an approach is conceivable for the realization of molecular layers since K. Blodgett and I. Langmuir have achieved the transfer of monolayers from water/air interface to solid surfaces[6, 7], thereby bequeathing their name to the method: the Langmuir-Blodgett (LB). Kuhn et al.[22] achieved in the 1970's nano-manipulation of molecules. To overcome the drawbacks of the LB method, J.Sagiv et al.[32] prepared SAMs by chemisorption, based on silane chemistry. With the same aim, Mallouk et al.[20] used crystalline chemistry. Since then, other techniques such as vapor deposition, nanolithography[34] have been developed for self-assembled monolayers (SAMs).

Except for the LB method, self-assembling molecules can be divided into three parts (fig. 23)[1].

Surface-active head group: This moiety will react with the surface. The choice of the head group depend on the surface used. Grafting can provide different chemical bonds (covalent, ionic, ...).

body group: Generally, the body is an alkyl, or derivatized-alkyl group. The ordering process is driven by van der Waals or electrostatic interaction.

Surface group: the choice of this group depends on the aim. This group gives its properties to the SAM (wettablility, reactivity, ...).

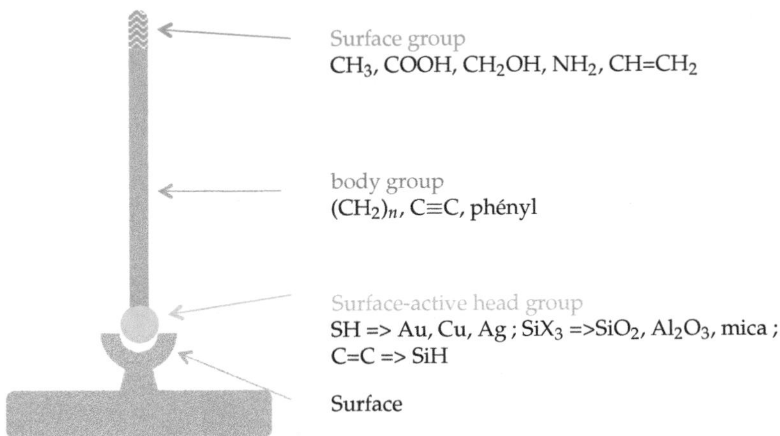

Figure 23. A schematic view of molecular structure [1]

Note that SAMs are more stable than LB films. Indeed, the latter are simply physisorbed, while self-assembled films, by their chemisorption, are more resistant to chemical attack and are more stable in temperature[45].

5.1.1. Self-assembly: interest

The interest of the self-assembly is in the control and the modification of the surface for specific applications. One can set the affinity of a surface with water or other solvents, such as protecting non-oxidized surfaces from water by making them hydrophobic[17]. SAMs also ensure biological compatibility for anchoring proteins[41] or conversely make bactericidal surface[38]. Note also the control of friction properties[25], the formation of very thin insulator[28, 46]. The combined use of these techniques with others, such as photolithography, allows also the organization of carbon nanotubes on a surface[15]. Furthermore, sequential self-assembly of SAM can be used to create 3D structures[3, 29, 30, 45].

5.1.2. Self-assembly: difficulties

Although the exothermic reaction[1] ($\sim$ 1-2 eV) between the surface-active group and the substrate promotes the use of a maximum of docking sites, forming a compact and orderly SAM on a large scale is not easy. Indeed, to enable the interchain interactions, it is first necessary that the molecules are close enough, requiring a high density of grafted molecules. Alkyl chains are tilted with respect to the normal to the surface by an angle (α), that depends on the recovery rate and defects[39]. Generally the higher the coverage, the smaller is α.

5.2. Self-assembly of organosilanes: generalities

The reaction of organosilane ($R_{(4-n)}$-Si-X_n; $n \in [1, 2, 3]$) derivatives with hydroxylated surface has attracted attention for several years. R is most often an alkyl chain, which may have different features, and X is an alkoxy or a chlorine. In the further study only organosilanes trifunctionalized ($n = 3$) will be used. They have the distinction of creating intermolecular bonds (fig. 24), which leads to the formation of an ordered molecular monolayer robust due to the crosslinking.

5.2.1. Grafting mechanism

The grafting process of organosilanes trifunctionalized takes place in four steps (fig. 24)[1].

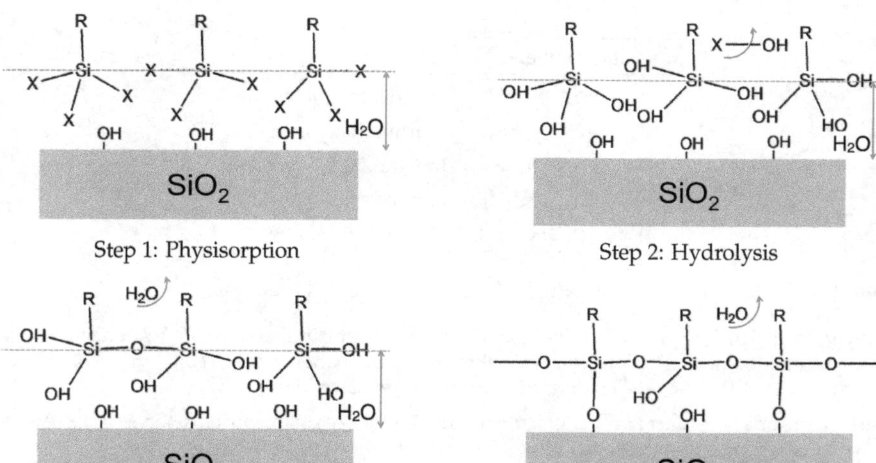

Figure 24. Different steps involved in the mechanism of SAM formation on a hydrated silica surface [1]

Step 1: Physisorption of molecules onto the adsorbed water monolayer(s) on the surface by the hydrophilic Si-R_3 moieties. Depending of the humidity rate there is one to three layer(s) of water on surface.

Step 2: Hydrolysis of the active-head group Si-R_3 by reaction with water.

Step 3: In plane reticulation: the surface-active head groups of two molecules react, and the reaction entails the formation of intermolecular siloxane bond(Si-O-Si). This reticulation is essential to the formation of close-packed and well-ordered SAMs.

Step 4: During this step molecule will react vita the surface forming siloxane bond (Si-O-Si).

During the various steps, the molecules, isolated or not, can diffuse on top of the water layer present on the surface. Through the interactions between neighboring molecules (hydrophobic body and active-surface group) under certain conditions leading to growth by islands, this mobility allows the molecules to compact enough for the transition from step 2 to step 3. Steps 3 and 4 are difficult to separate because they seem to occur almost simultaneously[8, 49].

5.2.2. Molecular structure influence

Each part of the molecule has an impact on the SAM growth[35]:

Surface-active head group: The kinetics of the monolayer formation is driven by the hydrolysis of the head group. Generally, the trifunctionalized organosilanes (R-Si-X_3) bear the same moieties X. Three moieties are particularly used: chlorine, methoxy and ethoxy. The table 9 present a classification by decreasing reactivity with SiO_2 as a function of the surface-active head group:

silane moieties	name
Si-Cl_3	trichlorosilane
Si-$(OCH_3)_3$	trimethoxysilane
Si-$(OCH_3)_2(OC_2H_5)_1$	dimethoxyethoxysilane
Si-$(OCH_3)_1(OC_2H_5)_2$	methoxydiethoxysilane
Si-$(OC_2H_5)_3$	triethoxysilane

Table 9. Classification of silanes moieties by decreasing reactivity with SiO_2 [40]

body group: The interchain van der Waals and electrostatic interactions are responsible of the final organization. The longer is the alkyl chain, the better is the organization. Note that beyond 18 carbons in the chain the SAMs are more disordered[1].

Surface group: This part of the molecule has few influence on the order. The use of large size groups can lead to steric hindrance.

5.3. Self-assembly of organosilanes: methodology

5.3.1. Cleaning process

Silicon substrates are cut from Si (100) wafers covered with native oxide. First, substrates are degreased in a sonicated chloroform bath (① fig. 25), and then dried under a nitrogen flow(② fig. 25). Substrates are then soaked into a piranha mixture (H_2SO_4, H_2O_2 30%; (v/v) 7:3, highly exothermic reaction, caution!) for 30 minutes at 150 °C (③ fig. 25) in order to remove any organic impurities from the surface and to increase the amount of hydroxyl moieties (OH) necessary for the grafting of silane heads. After that they are rinsed (④ fig. 25) with ultrapure deionized water (18 MΩ.cm) and quickly immersed into a beaker of de-ionized water(⑤ fig. 25).

5.3.2. Grafting process

Still in this beaker, substrates are introduced into a glove-box filled with nitrogen at 40% relative humidity in which silanization is performed. Substrates are dried under a nitrogen flow (① fig.26) and dipped into the silanization solution (② fig. 26), consisting of a mixture of hexadecane, carbon tetrachloride and the trichlorosilane at $10^{-2}M$ total concentration. This solution was beforehand thermalized on a thermostated plate during ~20 minutes at 11°C to which it is kept during all the silanization time of 1.5 to 2 hours. Then, samples are rinsed in a sonicated chloroform bath (④ fig. 26) and dried under a nitrogen flux (⑤ fig. 26).

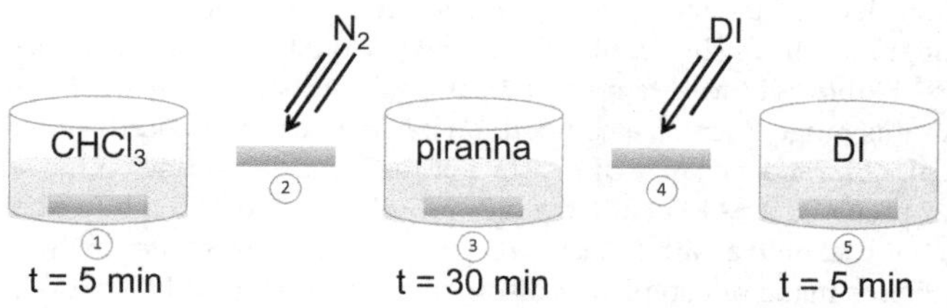

Figure 25. Cleaning process of silicon substrates

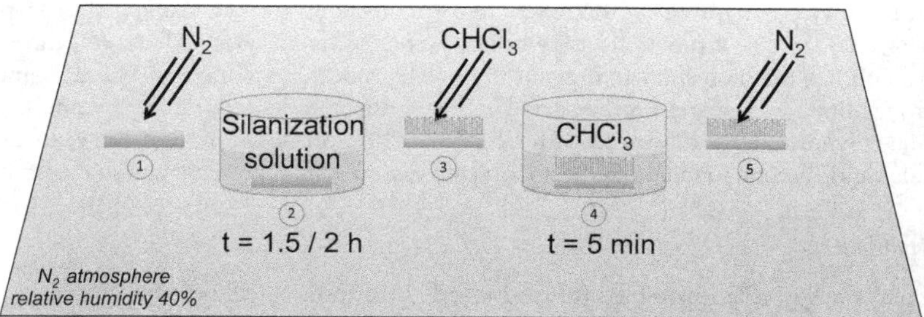

Figure 26. Silanization process

5.4. Self-assembly of organosilanes: critical parameters

In addition to the structure of the molecule and its density on the surface, the realization of a SAM from a solution involves other essential parameters to be controlled such as the nature of the solvent, the concentration, temperature and duration of the deposition which is conditioned by the kinetics reaction between the surface-active group and the surface.

5.4.1. Substrate cleaning

The silica surfaces are very sensitive to organic contaminants because of their high polarity. Simon Desbief, during his thesis at the IM2NP, tested different cleaning protocols (UV-ozone + HF, UV-ozone + HF + piranha, piranha ...) and their influence on the quality of an octadecyltrichlorosilane (OTS) layer. It appears that in the absence of a piranha treatment the layers obtained are very disorganized. The piranha removes most organic contaminants and makes the surface hydrophilic by promoting the formation of hydroxyl groups on the surface, i.e., attachment points of the molecules. On the other hand, results are not better when treatment is preceded by an attack of hydrofluoric acid and followed by a UV-ozonolysis.

5.4.2. Temperature

The grafting temperature is an important parameter. Contrary to what one might think, decreasing the grafting temperature favors the order of the layer. Brzoska et al.[9, 10] were the first to highlight this feature. They demonstrated the existence of a critical growth temperature (T_C) above which the monolayer is disordered. This temperature increases linearly with the length of the alkyl chain of the molecule, $T_C = Kn$ where n is the number of carbon forming the chain. Other studies have shown that the change of the type of growth does not occur at a temperature T_C but within a temperature range$[T_{C1};T_{C2}]$. When the temperature of the solution is in this range, there is a competition between ordered and disordered growth modes. Sung et al. observed a change of the state of the layer depending on the temperature of the bath. They prepared a SAM at a temperature $T<T_C$, then soaked it into a bath at $T>T_C$ and then plunged it back into a bath at $T<T_C$. It appears that in this sequence the layer grows in patches and then is made of completely disordered structure islands. It also highlights a pre-organization of the layer before grafting to the surface.

5.4.3. Solvent

Solvent polarity plays a role on the reactivity of molecules. In fact, the speed of steps 2 et 4 (fig. 24) increases as a function of the polarity of the solvent used. Moreover, the solvent had to solubilize the molecules to prevent their aggregation in solution and at the same time it has to facilitate its transfer on the surface. During the synthesis of binary monolayer[1], the molecule-solvent affinity allows to better control of the phase separation between the two molecules and thereby to facilitate the adsorption of a molecule with respect to the other.

5.4.4. Hydrometry

Water plays a key role during grafting. Indeed, studies have shown that the use of fully dehydrated substrates led to the formation of disordered layers, or conversely that the presence of water in the grafting solution activated the layer growth. Note that the presence of a big amount of water can cause polycondensation[11].

5.4.5. Concentration

The concentration of the grafting solution influences the kinetics of grafting. The more concentrated the solution is, the more molecules will interact with the hydroxyl groups of the surface. It should be noted that too high concentration may cause the formation and deposition of aggregates[11].

5.5. Shape analysis

It is thus interesting to statistically study the evolution in shape of such molecular islands according to various experimental parameters (time elapsed during the experiment, length of the molecule, temperature, etc.) in order to predict their further properties [36]. Images are obtained by microscopy with atomic force (AFM) and then analyzed. A model of identification card for a given island is then presented in Fig. 27. This card includes a number identifying the island, the location of its centroid, its perimeter and surface values, its feature values (CP: circularity parameter, $dmax$: maximal diameter, $hmax$: maximal height, $\bar{d}_{max}$: average diameter), its representation in 2D and 3D, and finally, the graphs of PN (point numbers from the boundary) and VH (average height variations). Then, a set of identification cards can be gathered in order to get statistical data.

---

[1] In this case there are 2 kinds of molecules in the silanization solution.

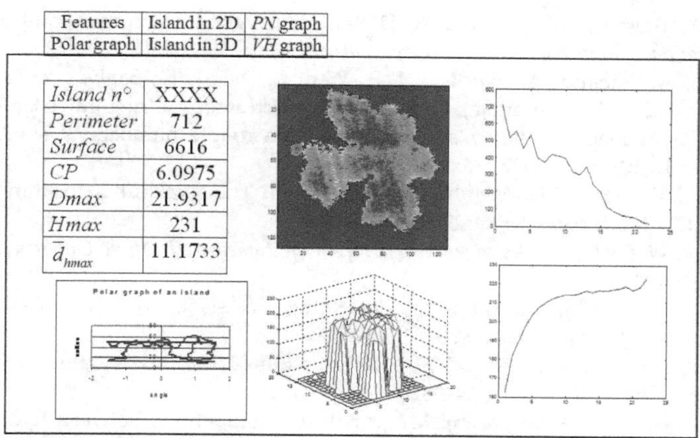

Features | Island in 2D | PN graph
Polar graph | Island in 3D | VH graph
Island n° XXXX
Perimeter 712
Surface 6616
CP 6.0975
Dmax 21.9317
Hmax 231
d hmax 11.1733
Polar graph of an island

Figure 27. Identification card of an island

6. Conclusion

In this chapter, a similarity parameter has been presented in order to compare shapes. Other parameters have been considered to determine shape features and their efficiency has been tested in real situations. Shape classification is achieved in an automatic way, that enables the processing of a huge amount of data, as well as statistical studies. The process has been illustrated in two different application contexts. For nuclear fuel particles, the estimation of the sphericity degree gives a mean of detection of what step of the fabrication is defective or deviating. This allows to characterize a sample of particles in order to determine if the batch they belong should be rejected or not. For molecular islands, the parameters define the identification card of each bounded surface. This leads to a statistical study in shape according to the time elapsed during the experiment.

Author details

F. Robert-Inacio and G. Delafosse
Institut Supérieur d'Electronique et du Numérique, IM2NP
CNRS, IM2NP (UMR 7334)

L. Patrone
Institut Supérieur d'Electronique et du Numérique, IM2NP
CNRS, IM2NP (UMR 7334)
Aix-Marseille Université, IM2NP, Maison des Technologies, Place Georges Pompidou, F-83000 Toulon, France

7. References

[1] Aswal, D. K., Lenfant, S., Guerin, D., Yakhmi, J. V. & Vuillaume, D. [2006]. Self assembled monolayers on silicon for molecular electronics, *Analytica chimica acta* 568(1-2): 84–108.

[2] Avis, D., Bremner, D. & Seidel, R. [1997]. How good are convex hull algorithms, *Computational Geometry: Theory and Applications* 7: 265–302.

[3] Baptiste, A., Gibaud, A., Bardeau, J. F., Wen, K., Maoz, R., Sagiv, J. & Ocko, B. M. [2002]. X-ray, Micro-Raman, and infrared spectroscopy structural characterization of Self-Assembled multilayer silane films with variable numbers of stacked layers, *Langmuir* 18(10): 3916–3922.

[4] Barber, C. B., Dobkin, D. & Huhdanpaa, H. [1996]. The quickhull algorithm for convex hulls, *ACM Trans. Math. Softw.* 22(4): 469–483.

[5] Blaschke, W. [1923]. *Vorlesungen über Differentialgeometrie. II: Affine Differentialgeometrie*, Springer, Berlin.

[6] Blodgett, K. B. [1935]. Films built by depositing successive monomolecular layers on a solid surface., *J. Amer. chem. Soc.* 57: 1007–1022.

[7] Blodgett, K. B. & Langmuir, I. [1937]. I.: Built-up films of barium stearate and their optical properties., *Physical Rev.* 51: 964–982.

[8] Brochier Salon, M. & Belgacem, M. N. [2010]. Competition between hydrolysis and condensation reactions of trialkoxysilanes, as a function of the amount of water and the nature of the organic group, *Colloids and Surfaces A: Physicochemical and Engineering Aspects* 366(1-3): 147–154.
URL: *http://www.sciencedirect.com/science/article/pii/S0927775710003389*

[9] Brzoska, J. B., Azouz, I. B. & Rondelez, F. [1994]. Silanization of solid substrates: A step toward reproducibility, *Langmuir* 10: 4367–4373.

[10] Brzoska, J. B., Shahidzadeh, N. & Rondelez, F. [1992]. Evidence of a transition temperature for the optimum deposition of grafted monolayer coatings, *Nature* 360: 719–721.

[11] Bunker, B. C., Carpick, R. W., Assink, R. A., Thomas, M. L., Hankins, M. G., Voigt, J. A., Sipola, D., de Boer, M. P. & Gulley, G. L. [2000]. The impact of solution agglomeration on the deposition of Self-Assembled monolayers, *Langmuir* 16(20): 7742–7751.
URL: *http://dx.doi.org/10.1021/la000502q*

[12] Charollais, F., Duhart, A. Felines, P., Guillermier, P. & C., P. [2004]. Influence of thermal treatment conditions on microstructure and composition of htr fuel kernel, *Ceramics Transactions of Am Ceram Soc* 168: 109–118.

[13] Charollais, F., Fonquernie, S., Perrais, C., Perez, M., Cellier, F. & Harbonnier, G. [2004]. Cea and areva r and d on htr fuel fabrication and presentation of the gaia experimental manufacturing line, *Proceedings of HTR2004*, Beijing, China.

[14] Charollais, F., Perez, M., Fonquernie, S., Ablitzer, C., Duhart, A., Perrais, C., Dugne, O., Guillermier, P. & Harbonnier, G. [2004]. Cea and areva fuel particles manufacturing and characterization r & d program, *Proceedings of Atalante2004*, Nimes, France.

[15] Choi, K. [2000]. Controlled deposition of carbon nanotubes on a patterned substrate, *Surface Science* 462(1-3): 195–202.

[16] Danielsson, P. [1980]. Euclidean distance mapping, *CVGIP* 14: 227–248.

[17] Duchene, S., Davidovits, N. & Millasseau, F. [n.d.]. *Brevet Français numéro 96 0.1319* .

[18] Gauthier, J., Lecomte, M., Brinkmann, G. & B., B. C. [2004]. Antares: the htr/vhtr project at framatome anp, *Proceedings of 2nd int meeting on HTR Techn 2004*.

[19] Grünbaum, B. [1963]. Measures of symmetry for convex sets, *Proc. Symp. Pure Math.*, Vol. 7, pp. 233–270.

[20] H.Lee, Kepley, L., Hong, H. & Mallouk, T. [1987]. Inorganic analogues of Langmuir-Blodgett films: Adsorption of ordered zirconium 1,10-decanebisphosphonate multilayers on silicon., *J Am. Chem. Soc.* 110: 618–620.
[21] Ho, C. & Chen, L. [1995]. A fast ellipse/circle detector using geometric symmetry, *Pattern Recognition* 28(1): 117–124.
[22] Inacker, O., Kuhn, H., Mobiusand, D. & Debuch, G. [1976]. Manipulation in molecular dimensions., *Z. physik. Chem. Neue Folge* 51: 337–360.
[23] Kazhdan, M., Chazelle, B., Dobkin, D., Finkelstein, A. & Funkhouser, T. [2002]. A reflective symmetry descriptor, *Proc. European Conference on Computer Vision (ECCV)*.
[24] Kazhdan, M., Funkhouser, T. & Rusinkiewicz, S. [2004]. *Symmetry descriptors and 3D shape matching*, R. Scopigno and D. Zorin editors.
[25] Kim, H. I., Graupe, M., Oloba, O., Koini, T., Imaduddin, S., Lee, T. R. & Perry, S. S. [1999]. Molecularly specific studies of the frictional properties of monolayer film : a systematic comparaison of cf_3, $(ch_3)_2ch$, and ch_3 terminated films., *Langmuir* 15(9): 3179–3185.
[26] Labouré, M., Jourlin, M., Fillère, I., Becker, J. & Frénéa, F. [1996]. Isoperimetric inequalities and shape parameters, *Acta Stereologica* 15/1: 65–70.
[27] Languille, A. [2002]. High temperature reactor fuel technology programme in europe, *Proceedings of HTR-TN 2002*, Petten, Holland.
[28] Mann, B. & Kuhn, H. [1971]. Tunneling through fatty acid salt monolayers, *Journal of Applied Physics* 42: 4398.
[29] Maoz, R., Matlis, S., DiMasi, E., Ocko, B. M. & Sagiv, J. [1996]. Self-replicating amphiphilic monolayers, *Nature* 384: 150–153.
URL: *http://www.nature.com/nature/journal/v384/n6605/pdf/384150a0.pdf*
[30] Maoz, R., Sagiv, J., Degenhardt, D., Miihwald, H. & Quint, P. [1995]. Hydrogen-bonded multilayers of self-assembling silanes: structure elucidation by combined fourier transform infra-red spectroscopy and x-ray scattering techniques, *Supramolecular Science* 2(1): 9–24.
[31] Marola, G. [1989]. On the detection of the axis of symmetry or almost symmetric planar images, *IEEE PAMI* 11: 104–107.
[32] Netzer, L., Iscovici, R. & Sagiv, J. [1983]. Adsorbed monolayers versus langmuir-blodgett monolayers- why and how ? 1: From monolayer to multilayer by adsorption., *Thin Solid Films* 99: 235–241.
[33] Phelip, M. [2003]. First results from htr-f&f1 projects : Htr fuel technology, *Proceedings of International Congress on Advanced nuclear Power Plants ICAPP'03*, Cordoba, Spain.
[34] Piner, R. D., Zhu, J., Xu, F., Hong, S. & Mirkin, C. A. [1999]. "Dip-Pen" nanolithography, *Science* 283(5402): 661 –663.
[35] Rauscher, H. [2001]. The interaction of silanes with silicon single crystal surfaces: microscopic processes and structures, *Surface Science Reports* 42(6-8): 207–328.
[36] Robert-Inacio, F. [2008]. Shape studies in 2d and 3d on molecular islands of carbon chains self-assembled on silicon in nanoelectronics, *European Physical Journal: Applied Physics* 41/1: 53–68.
[37] Robert-Inacio, F., Boschet, C., Charollais, F. & Cellier, F. [2006]. Polar studies of the sphericity degree of v/htr nuclear fuel particles, *Materials Characterization* 56(4-5): 266–273.

[38] Rondelez, F. & Bezou, P. [1999]. Des surfaces antibactériennes et autostériles, *L'actualité chimique* 10: 4–8.

[39] Schreiber, F. [2000]. Structure and growth of self-assembling monolayers, *Progress in Surface Science* 65(5-8): 151–257.
URL: *http://www.sciencedirect.com/science/article/pii/S0079681600000241*

[40] Simon, A. [2002]. *Intérêt de la microscopie de force atomique sur la bio fonctionnalisation de matériaux : caractérisation du greffage et de l'adhésion cellulaire.*, PhD thesis, Université Bordeaux I.

[41] Sugimura, H. & Nakagiri, N. [1997]. Nanoscopic surface architecture based on scanning probe electrochemistry and molecular Self-Assembly, *J. Am. Chem. Soc.* 119(39): 9226–9229.

[42] Sun, C. [1995]. Symmetry detection using gradient information, *Pattern Recognition Letters* 16: 987–996.

[43] Sun, C. & Si, D. [1999]. Fast reflectional symmetry detection using orientation histograms, *Real-time Imaging* 5: 63–74.

[44] Süss, W. [1950]. Ueber eibereiche mit mittelpunkt, *Math-Phys* Semesterber 1: 273–287.

[45] Ulman, A. [1991]. *An Introduction to ULTRATHIN ORGANIC FILMS From Langmuir-Blodgett to Self-Assembly*, ACADEMIC PRESS, Inc, London.

[46] Vuillaume, D., Boulas, C., Collet, J., Davidovits, J. V. & Rondelez, F. [1996]. Organic insulating films of nanometer thicknesses, *Applied Physics Letters* 69: 1646.

[47] Wang, H. & Suter, D. [2003]. Using symmetry in robust model fitting, *Pattern Recognition Letters* 24: 2953–2966.

[48] Wolter, J., Woo, T. & Volz, R. [1985]. Optimal algorithms for symmetry detection in two and three dimensions, *The Visual Computer* 1: 37–48.

[49] Yang, Y., Bittner, A. M., Baldelli, S. & Kern, K. [2008]. Study of self-assembled triethoxysilane thin films made by casting neat reagents in ambient atmosphere, *Thin Solid Films* 516(12): 3948–3956.
URL: *http://www.sciencedirect.com/science/article/pii/S0040609007013685*

Chapter 8

Microfluidic Devices Fabrication for Bioelectrokinetic System Applications

Nurul Amziah Md Yunus

Additional information is available at the end of the chapter

1. Introduction

The research and development of microfluidic technologies has increased tremendously over the past twenty years. The technology allows designers to create small, portable, robust, low cost and easy-to-use diagnostic instruments that offer high levels of capability and versatility. Microfluidic systems will decrease reagent consumption and reduce cost per analysis. It also reduces analysis time and provides better controllable process parameters in chemical reactions.

Microfluidic is also known as miniaturised total (chemical) analysis system, (μTAS). It is a generic term for a small system or device (microfluidic) designed to perform one or more chemical/biochemical processes. The early stage of microfluidics has been dominated by the development of microflow sensors, micropumps and microvalves in late 1980s and the beginning of 1990s and the concept has been introduced by Manz and co-workers from Imperial College London [1]. Since then, chip-based analytical systems have been rapidly applied to a variety of fields such as separation science, chemical production, DNA analysis, medical diagnostics and environmental analysis [2, 3]. Examples of microfluidic devices are shown in Fig. 1 (a)-(b).

Surprisingly, many fabrication methods used for making a microfluidic device are based on silicon microelectronics fabrication industry, with one of the first devices on silicon by Terry [a gas chromatographic air analyser in the late 70s, but little study was done about producing one again at that time. Interest in this type of technology was revived in the 90s with current microfluidic devices fabricated not only from silicon based but to a range of polymers, glass wafers and foils. This technology is used to build miniature devices as shown in Fig. 1(c) with packaging that could perform as a conventional "big bench-top" such as shown in Fig. 1 (d).

Thus, the challenge has been to create such as medical, chemical, biological devices that are capable of doing a fast and accurate analysis of the reagent and has the capacity to be green technology products in the future. With the recent advances in the synthesis and the characterization of size-selected particles in colloids (submicron and nanometer range) such as blood itself, an investigation on their physical and chemical properties has been made possible [5]. In microfluidic, a fabrication of it devices with integrated microchannels and microelectrodes of dimensions are made comparable to biological cells or particles size. Additionally, if there is a capability of producing small-scale devices, it will allow the development of entirely novel experiments, which currently is rapidly under study.

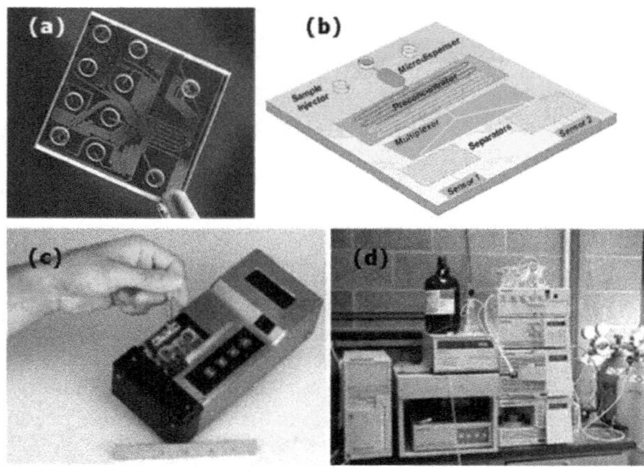

Figure 1. (a) Microfluidics Device –(image is taken from Agilent Technologies website) (b) Possible analysis on the microfluidic device with integrated microfluidic with sample injector, microdispenser, preconcentrator, multiplexor, separator and sensor (c) Handheld microfluidic in the proper packaging (d) Example of a big benchtop system –the conventional version of chemical/biological analytical technique.

Microchannels are very important features for microfluidic systems. There are many ways to fabricate the microchannel, such as by using the silicon based organic polymer, polydimethylsiloxane (PDMS) or the femtosecond laser micromachining and inscription or using deep reactive ion etching (DRIE). The method emphasized in this Chapter will be the fabrication of microfluidic microchannel using the laminate sheet, dry film resist (DFR), for example, the SY series from Elga Europe [7]. This laminate sheet is a biocompatible material and it adheres easily on the co-planar substrate, i.e glass or silicon substrate. It is cheap, simpler and easier to use.

The work is also concerned on the application of electrokinetic technique to bioelectrokinetic monitoring system. Electrokinetic is defined as the study of fluid or particle motion in electric fields. It generally works in direct current, DC and alternating current, AC. Electrokinetics include electrophoresis (EP), dielectrophoresis (DEP), electro-osmosis (EO), electrorotation (ROT) and electroorientation [6]. These electrokinetic processes are able to manipulate, concentrate, collect, mix and separate different types of bioparticles, biomolecules, organic and inorganic materials.

2. The physics of microfluidic systems

In this section, an overview of the physics of microfluidic will be highlighted because at the microscale, there are many different forces which become significant over forces experienced in everyday life. It is begun with the motion of fluid described by Navier-Stokes equation named after Claude-Louis Navier (1785-1836), a French engineer and physicist and George Gabriel Stokes (1819-1903), a mathematician and also a physicist. The equation was derived from conservation of momentum arguments [8]. This equation arises from applying Newton's second law to fluid motion, together with the assumption that the fluid stress is the sum of a constant viscous term (proportional to the gradient of velocity), plus a pressure term [9]. For an incompressible Newtonian fluid (named after Isaac Newton a mathematician and physicist (1642-1727) [10]) the equation is [11, 12]:

$$\rho_m \frac{\partial \mathbf{u}}{\partial t} + \rho_m (\mathbf{u} \bullet \nabla) \mathbf{u} = -\nabla p + \eta \nabla^2 \mathbf{u} + \mathbf{f} \tag{1}$$

where ρ is the mass density, $\mathbf{u}$ is the velocity of fluid, p is the pressure, η is the viscosity and $\mathbf{f}$ is the total applied body force. The ratio of the inertial term to the viscous term can be determined from the coefficients in equation which gives a factor referred to as the Reynolds' number (named after Osborne Reynolds (1842–1912)) [13-15] shown in equation (2):

$$\mathrm{Re} = \frac{\rho_m u_o l_o}{\eta} \tag{2}$$

'Re' stands for Reynolds' number, u_o is the magnitude of a typical velocity and l_o is a length scale. It is a dimensionless number and can be estimated as follows: Assuming that typical working fluid is water with typical velocities u_o of 10^{-6} to 10^{-2} m/s. Typical channel radii of 10^{-4} m, $\rho_m = 10^3 kgm^{-3}$ and $\eta = 10^{-3} kgm^{-1}s^{-1}$. For fluids in microfluidic devices, which have small Reynolds number (Re <<1), their behaviour is dominated by viscous forces over inertial forces and the resulting flows are linear and the fluid is considered to move in laminar sheets. If inertial forces are much bigger, Re>>1, turbulent flow will occur. The low values of the Reynolds number in the microfluidic microsystem indicate that, if the fluid is moving under an applied pressure or force, which is suddenly removed, the fluid flow will stop immediately.

2.1. Laminar flow

In microfluidic microsystem, the microchannel is fabricated to guide the fluid flow through the device. This channel can be in any size in the range of less than 5μm to 100s μm and the flow is always laminar. Laminar by definition is a layer. Laminar flow is a condition whereby the velocity of a particle in a fluid stream is not a random function of time thus the fluid flow follows streamlines [16]. One of the effects of laminar flow is that, two or more streams flowing in contact with each other will not be mixed except by diffusion [17].

The fluid in steady state flow is defined as Poiseuille flow (named after Jean Louis Marie Poiseuille 1797-1869) a French physician and physiologist) [18]. The flow establishes a parabolic flow profile after some distance from the entry point of the fluid into the channel. The velocity is zero at the walls and reaches maximum at the centre of the channel.

2.2. Viscous drag force

When there is particle in fluid, it will experience a viscous drag force due to the action of the fluid on the particles. Viscous drag is the force that resists the movement of a particle through the fluid. The force is proportional to but opposite the relative velocity of the particles. For a sphere particle of radius a in a fluid of viscosity η, viscous drag force is given by Stokes's law [11] in equation (3).

$$\mathbf{F}_{Viscous} = 6\pi\eta a\mathbf{v} \tag{3}$$

$\mathbf{v}$ is the velocity of the particle. For a constant applied force the particle will eventually reach a terminal velocity, which makes it, does not accelerating. If the fluid is at rest, the terminal velocity is simply proportional to the applied force. If the fluid is moving, the terminal velocity experienced by the particle depends on the velocity of the fluid.

For a sphere particle, the friction factor is given by equation (4). The constant *f*, which refers to the friction factor, depends on a range of particle parameter such as size, shape and surface characteristics.

$$f = 6\pi\eta a \tag{4}$$

3. Fabrication of microfluidic process

When the microfluidic is introduced in the early 1990s, a glass substrate is used as a material for the substrate for most microfluidic devices [19-21].The advantages of glass in relation to the other materials are it is chemically inert to most liquid and gases, hydrophilic and optically 100% clear. There are three types of glass that are commonly used in the laboratory. There are consisting of borosilicate, soda lime and fused silica glass. In this work, the borosilicate glass or pyrex glass, which is the type of the microscopic glass slide, is used as the substrate. It can resist strong acids, saline solutions, chlorine, bromine, iodine, and strong oxidizing and corrosive chemicals. This glass can be produced in a mass product since it is not expensive.

The other reason glass slide is used as a substrate, is that glass is hydrophilic, in which it attracts and holds moisture. Most plastics, in comparison, are hydrophobic and need treatment to become hydrophilic. Glass has material properties that are stable in time and it is thus preferable for applications in which devices are used extensively, such as in high throughput screening. It is a reliable shelf life. It is non-porous, implying that small molecules will not be able to diffuse into it.

3.1. Glass as a substrate

To start utilize glass for microfluidic device, we could cut it into pieces, of dimension for example 2 cm by 1.5 cm, and we call it as fragmented chip glasses. These glasses will go through the cleaning process. They were placed in the glass holder filled with 95% pure water and 5% of soap. Then this holder is placed in the ultrasonic bath (Ultrawave Ltd) for washing away all the fouls with the aid of the beaker filled with DI water for 15 minutes. After cleaning, the glasses are thoroughly rinsed with DI water followed by acetone, methanol and iso-propanol (IPA), the RGB solvents, and blow-dried with nitrogen.

The glasses are placed in the oven for dehydration purposes at the temperature of 200°C for 2 hours minimum after the cleaning process is completed. This step is done to make sure that the glasses used are dried totally so that attachment of laminate sheet mentioned earlier in the introduction on these glasses will be excellent.

3.2. The device

Manz and co-workers introduced photolithographic techniques to microfabricated electrophoretic separation channels [19]. Lithography uses photoresist materials to cover areas on the wafer that will not be subjected to material deposition or removal.

There are two types of photoresist materials, namely, negative and positive photoresists. Negative photoresists are those that will become less soluble in the developer solution when they are exposed to light, forming negative images of the mask patterns on the wafer (the substrate) while positive photoresists are those that will become more soluble in the developer when they are put under the same exposure of light, forming positive images of the mask patterns on the wafer.

Advances in micro and nano fabrication techniques allow small microelectrodes of order of 1μm in size and smaller with selective materials to be manufactured with relative ease. Therefore microfluidic device has been designed and fabricated by taking into consideration the material used for making the microelectrode so that only small amount of voltage needed to gain high current. Thus, this microelectrode structures can provide sufficient force to manipulate particles without requiring high voltage signal generators. This will reduce joule heating in the microfluidic device. Although, it is found that if the microelectrode gap and chamber height is made smaller and lower respectively can impinge on the joule heating problem when higher conductivity medium is employed especially by biologist in biology application, but with selective material for microelectrode i.e biocompatible, it will overcome that problem.

The microelectrodes can be fabricated on approximately 10 cm diameter, (4 inch), 1 mm thick glass wafers or substrates. The total chip that can fit the 4 inch wafer is about 14 chips.

This is depending on the size of the device/chip designed by the engineer for a particular function. The microelectrode will be deposited on the glass wafer by using the electron beam lithography. This technique will ensure the layer of microelectrode is thinner and the electrode width and gap are precisely small.

The microelectrodes are fabricated using gold, Au. The gold layer is essential because of its low resistance and its biocompatible nature, but unfortunately it does not adhere well to the glass surface. Thus titanium, Ti on the other hand, adheres well on the glass substrate and was used to improve gold adhesion. The palladium, Pd layer acts as a diffusion barrier between the titanium, Ti and the gold, Au. The other metal that can promote adhesion to gold, Au and platinum, Pt is chromium, Cr. However Cr is known to diffuse into the overlying gold, Au over time more quickly than titanium, Ti.

Microelectrodes on the glass were patterned using photolithography and ion beam milling over titanium, Ti, seed layer. Pt or Au microelectrodes are chosen because they are highly conductive, chemically unreactive, and able to resist tarnishing and corrosion, which then would allow the maximum current flow through it and hence, greater force or energy could be produced. They are also biocompatible because of its quality of not having toxic or injurious effects on biological systems. Some evaporated metal layers for microelectrode are 10 nm titanium, Ti, 10 nm palladium, Pd, and 100 nm gold, Au. For example, in this work, the width of the wire of the microelectrode is 20 μm and it is made of three layers of metal consisting 30 nm Ti, 100 nm Au and 30 nm Ti. The top layer of Ti is to reduce the effects of corrosion at low frequency and high potentials [22]. The electron-beam lithography technique is used to fabricate smaller size of microelectrode ranging 1-5 μm and gap sizes ranging from 1-10μm and its reproducibly.

3.3. Laminate

The next step is to perform the channel on these glasses. The fresh laminate sheet, dry film resist (DFR) of 50 μm is taken out from the fridge. The purpose of keeping the laminate in cold place is to lengthen the freshness/lifetime/dryness/unwanted crosslink. It is folded in black plastic cover to prevent exposure to or penetration of the UV light. A layer at one side of the laminate is the protective film, the polyethylene (PE) layer, is then peeled off. The photoresist is then applied on the substrate, which is placed on the cardboard as shown in Fig. 1. The substrates are put through the hot-roll laminator with speed 1 m/min and at temperature 100˚C -110˚C. The protective polyester (PET) layer should remain on top of the laminate when the exposure process is performed. This layer is to be removed once the processes of developing and rinsing are followed.

3.4. Adhesive bonding

Our procedure for bonding the fragmented chips uses a technique called adhesive bonding. There are two appliances that can be used for adhesive bonding process, the hot press bonder and the oven. For the hot press bonder, the temperature used must be ramped up by 0.83°C/min, from 150°C to 200°C. The length of time of the first temperature, 150°C is set for

30 minutes while the second temperature 200°C is set for 1 hour. After the course, the temperature is ramped down slowly, with 1°C/min until it settled to the room temperature. This process will take approximately 6 hours to be completed.

For bonding in the oven, the chip will be pressed between several standard size microscope glass slides (rigid plates) with additional big black paper clamps to be used for clamping as shown in Fig. 2. There is other presser that we can use i.e the ceramic type presser. This cheap procedure follows the ideas from Pan et al. [23]. The process is started by setting the temperature ramped up 0.83°C/min to 150°C for 30 minutes, and then it is ramped down 1°C/min to the room temperature. The step is followed by setting up back the temperature 0.83°C/min to 200°C for two hours and then the temperature is ramped down 0.83°C/min to the room temperature. This process will take longer time as compared to the hot press bonder [7].

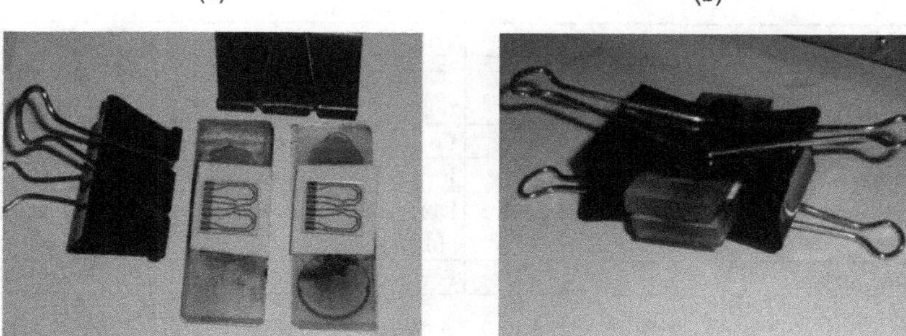

Figure 2. (a) Picture of the substrates that are ready to be placed in between the set of glass slides with Teflon sheets as the absorber/suspension to give room for the expansion and contraction when the temperature rise or fall and (b) they will be clamped with two big black paper clamps which are used for clamping the glasses at its place.

Subsequently, the channel made will be drilled at specific inlets and outlets (marked) to form the holes for the fluid to flow into and out from the channel. The test fluids are flowed through the channel to check for any leakage and clogging before the device get fully utilised [24]. Sometimes the bonded channel are tested using fluorescent dyes to confirm a good seal [25] otherwise the UV glue will be used if the channel has leakage. This UV glue is used to cover the high probability of leak area at the side of the two bonded substrates. The glue will flow by capillary force in between the substrates. And, when the glue has covered the whole area around the channel wall, the chip will then be exposed under the UV light for about 40s. As a result, the close channel will be formed successfully.

3.5. Channel fabrication

The mask design can be created using L-Edit/CleWin or any CAD drawing tools and this design can be transferred in a high-resolution transparency/acetate mask using a high-resolution printer suggested to be from 3386 dpi (dots per inch) to 128000 dpi. Fig. 3 below shows the ready made acetate mask of the channel. The kind of this mask is preferred and the advantages of it are stated in Table 1.

Illuminated areas will be remained on the glass during development process because we will be using negative photoresist. The exposure time and the developing time are then recorded. This process, of course, requires a very clean substrate (wafer) so that no airborne debris or dust imbedded onto mask when the mask makes contact with the wafer. The imbedded particle could cause permanent damage to the mask and result in defects on the glass wafer with each succeeding exposure.

Parameter	Mask
Pattern	1:1 mask-substrate
Critical dimension	Easy to pattern micron dimensions on mask and substrate
Exposure Field	Entire substrate
Mask technology	Mask has same dimensions as substrate -a rapid prototype
Throughput	Potentially higher
Die (chip/device) alignment and focus	Individual die alignment and focus
Defect density	Defects are not repeated multiple times on a substrate
Surface flatness	For overall global focus and alignment

Table 1. The advantages of using pattern acetate mask-substrate transfer.

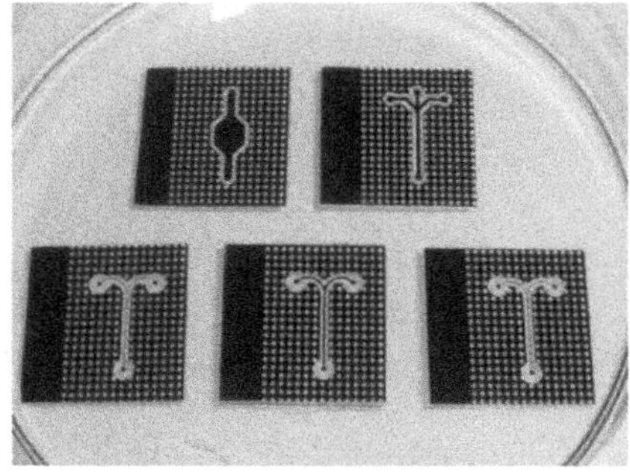

Figure 3. Acetate mask of the channel, with white area is the part where the photoresist is to be remained on the substrate and black area is the part where the photoresist is to be removed after the patterning process.

3.6. Bonding

Prior to bonding, the devices with microfluidic channel are first treated with plasma asher, (Oxford Instruments Plasmalab 80 Plus System) where it is used for ashing, etching and cleaning the surface of polymer, the microchannel. It will remove the surface contamination and prevent any contamination from interfering the adhesion especially during bonding process. It makes the surface more hydrophilic and thus enhances the adhesive transfer. In other words, plasma treatment will improve polymer analysis, wettability problems, painting/coating of plastics, gluing results and clean surfaces from carbon, grease and oil. The interactions of plasma with a polymer surface can be divided into four general categories as shown in block diagram in Fig. 4.

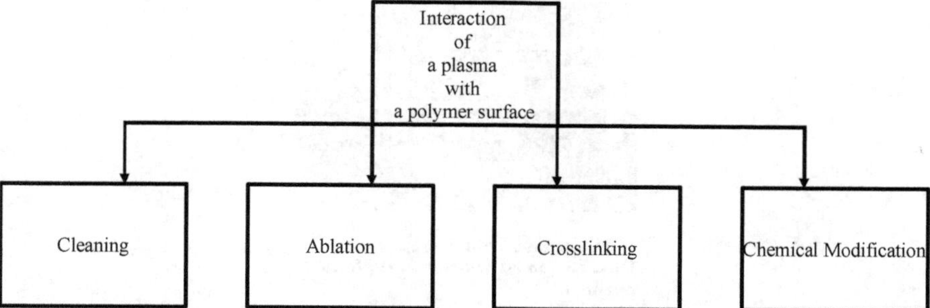

Figure 4. Four main interactions of plasma with a polymer surface. Cleaning, Ablation, Crosslinking and Chemical Modification.

A surface cleaning is the removal of organic contamination. Ablation is to remove material by micro-etching mainly to increase surface area and it is also used to remove a weak boundary layer. A cross-linking or branching is used to strengthen the surface cohesively, and surface chemistry modification is to improve chemical and physical interactions at the bonding inter-phase. The setting of the equipment is shown in Table 2.

Setting	**Condition**
O_2	10 sccm (process gas out)
Pressure	50 (set)
Forward Pressure	100 (RF)
Chiller	20 degree

Table 2. Sample using plasma asher for 2 minutes with its settings and conditions.

These microchannels are sandwiched between two glass substrates to form a closed microchannel for microfluidic systems applications such as a micropump, micromixer and separation systems. For example, the microchannel size is 500 μm wide and the step by step of forming the channel has been discussed in [7]. First, the microchannel is placed at one side of the glass substrate with microelectrodes array on it. The other side of the glass substrate will only with another microchannel on it without microelectrode as shown in Fig. 4. The disadvantage of this configuration with one side of microelectrodes array and two layer of laminates is that the electrokinetic force is weaker due to small gradient of electric field strength generated and it will be difficult to be used to control the bioparticle motion [26].

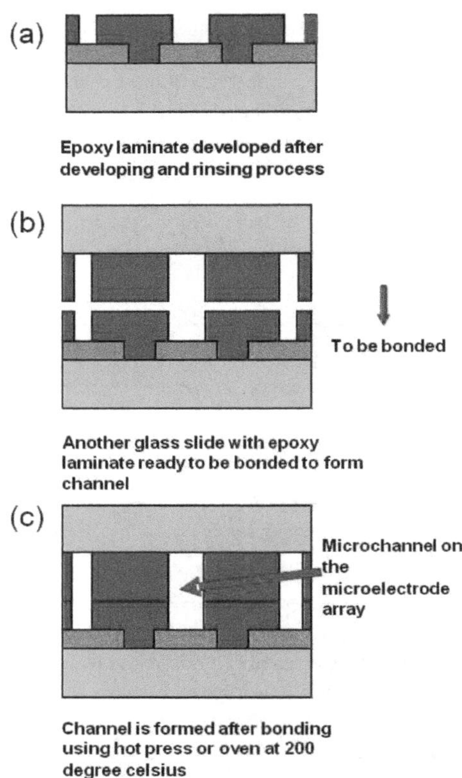
(a)
Epoxy laminate developed after developing and rinsing process
(b)
To be bonded
Another glass slide with epoxy laminate ready to be bonded to form channel
(c)
Microchannel on the microelectrode array
Channel is formed after bonding using hot press or oven at 200 degree celsius

Figure 5. (a) The developing and rinsing process of laminate sheet deposited on the substrate with microelectrodes array (b) The two channels from two glass substrates are aligned and then bonded and (c) hence form the close channel after placing them in the hot press or oven at 200°C.

The electric field strength of the microfluidic device is further increased and this is done by placing the microelectrodes array on the top and the bottom of the glass substrate and with only one layer of laminate as the microchannel. The microfluidic device for bioelectrokinetic working much better with this arrangement, because the gradient of the electric field generated becomes stronger [27]. The thickness or the height of the channel is reduced from 50 μm to approximately 35-37 μm since only one laminate sheet is used [28]. The channel height is reduced compromising the desired electrokinetic operation. For example, we can use it as a separation device to separate a mixture of submicron particles.

Nevertheless, there are times where the bonding process has not been successful. If more than one device is intended to be bonded at the same time using the hot press bonder as shown in Fig. 6, mainly when they are collected to be bonded together, the thickness of the device must really be considered. Otherwise the results shown will be a part of the microfluidic channel area bonded and a part will not be bonded. Fig. 7 shows the result after using the hot press bonder with collection of devices having different thicknesses to be bonded and assembled in one go.

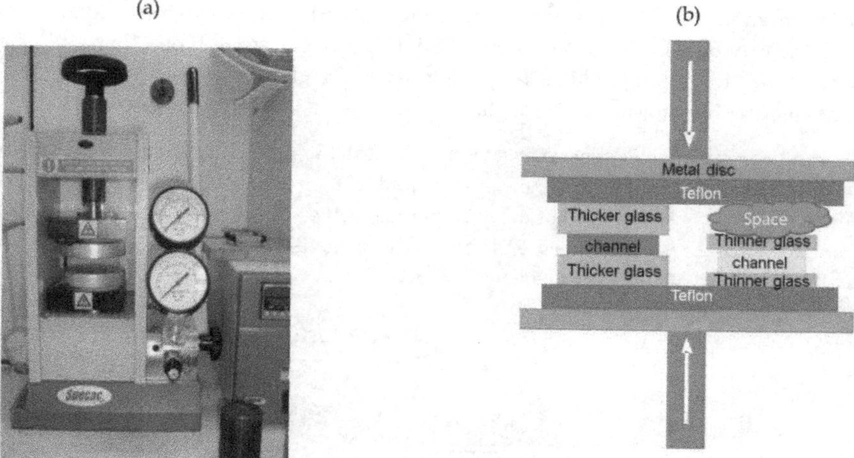

Figure 6. (a) The hot press bonder (b) Illustration of two uneven sized/height devices bonded together in the hot press which will cause the non-bonded area.

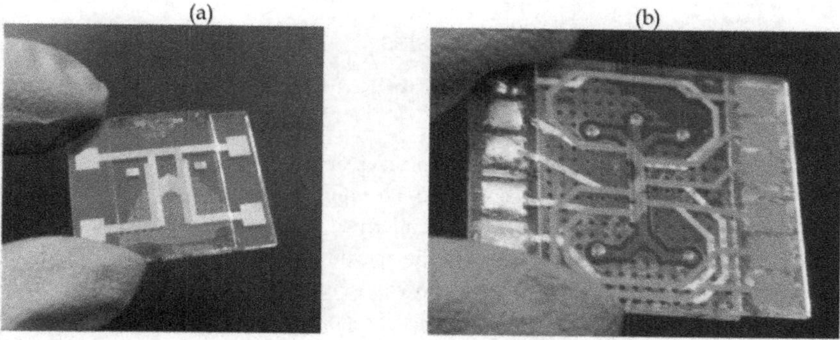

Figure 7. (a) The non-bonded area of the device approximately 2 mm thick was shown at the bottom part of the device with light shaded area (b) the non-bonded area of the device approximately 1.4 mm thick is shown at the right hand top corner near the right most top inlet.

Furthermore, high internal pressures could be developed in the hot press bonder as the fabricated channels were compressed between two contact glasses (wafers), and this might create non-bonding forces as well [29]. These occurrences are not preferred because, the channel will not be formed properly and if the aperture is drilled, there will be a leakage. The optimization of the bonding process for fragmented device is still under investigation.

3.7. Drilling

It is important to exercise a great caution when handling or drilling apertures into the device since the device is small and made from glass. The machine used to drill aperture on the device is from ProxxonGmbH i.e the main adapter NG 5/E with bench drill TBM 220 (220-240V). The drill bit used is a solid carbide drill, DCSPADEG, spade type, 1 mm, 60º point from Drill Service (Horley) Ltd. It is specially made for glass substrate. The sharp point helps to reduce chipping at breakthrough.

The process of drilling will generate heat when the drill bit is spinning and it may break the device. Therefore to reduce heat, the chip is placed in the petri dish filled with water. The other factor to be considered is that the drill bit must be very sharp to avoid fractal crack on the surface of the glass. The speed and the height of the drill bit must also be controlled. Here the speed of drilling is approximately 600rpm.

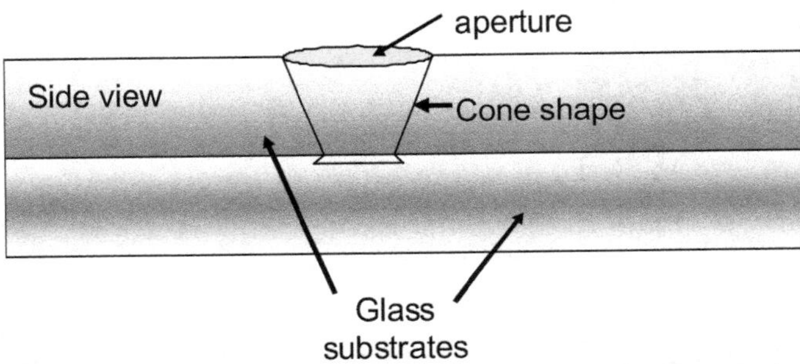

Figure 8. The chip is tilted (side view) to see the shape of the cone has been formed on the both side near the drilled aperture.

The process of drilling starts after the correct place of aperture is marked. In order to make sure that the aperture has already been drilled, the chip is tilted to the side under the light to check whether the shape of the cone has been formed on both glass sides near the drilled aperture as shown in Fig.8. The geometry of the aperture depends on the shape of the drill bit. The device could crack at apertures as shown in Fig. 9(a) and (b). The chip might also break into pieces if the apertures are not drilled properly as shown in Fig. 9(c)-(e) and to overcome this occurence, a temporary tape could be placed on top of the device to prevent leakage as shown in Fig. 9(f).

Since the apertures are drilled using a machine and we are the one who controlled the speed, we might get a different aperture diameter in the end. This will result some differences on the input or output resistance. Once this occurred, there will be a problem if one is working to balance the sheath flow or to sort the particles at the output. In other word, a robotic and controllable driller for microfluidic device should be designed to avoid differences in input and output resistance of the device.

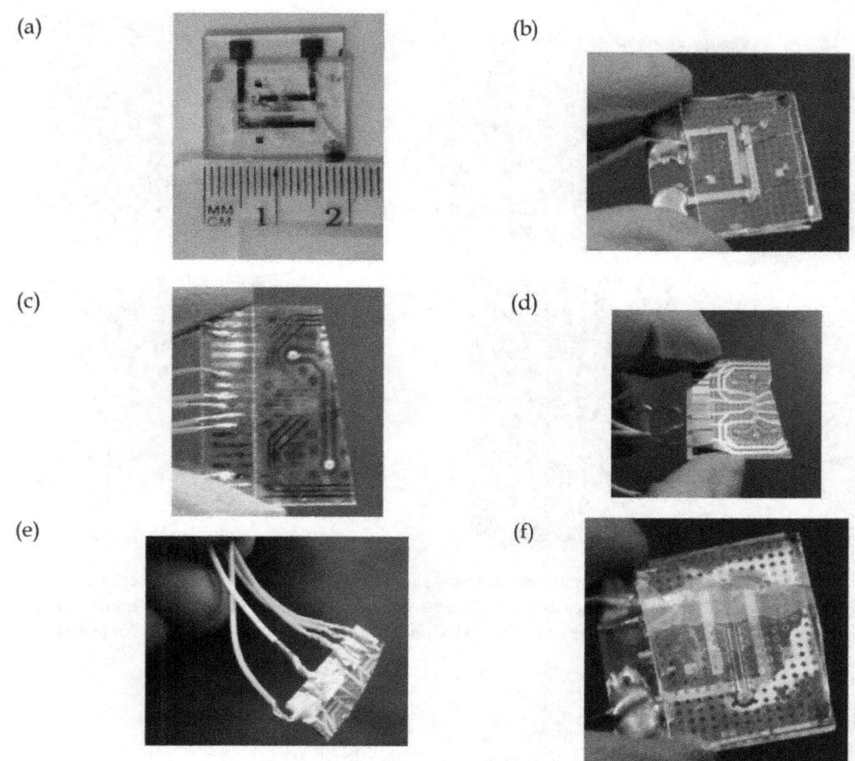

Figure 9. (a) and (b) show the devices crack because of the drilling process. (c), (d) and (e) show the devices break into pieces when there is severe crack occur during drilling process. This breakage will cause trouble during fitting the chip into the holder or handling/preparation of the chip before the experiment. (f) The device is plastered using the tape to cover the other side of the substrate when the drilling process has accidentally passed through the other side of the substrate making the hole throughout the substrates.

3.8. Channel formation

Finally, the channel is produced after laminating and drilling apertures are completed. This close channel is shown in Fig. 10. Fig. 10(a) shows the earlier device tested however has residues in it. Fig. 10(b) shows the successful bonded device under the temperature of 200°C with the length of heating is 2 hours. Fig. 10(c) shows the successful bonded device under temperature 200°C with length of heating of 4 hours. Notice that the image of the device is darker in colour when it is heated longer. Fig. 10(d) shows a device that can be examined by running dyed water into it. This is done to observe if there is any leakage by using the syringe connected with the silicon tube. The bonded device is then tested with chemical resistance by continuously flow the chemical through the channel.

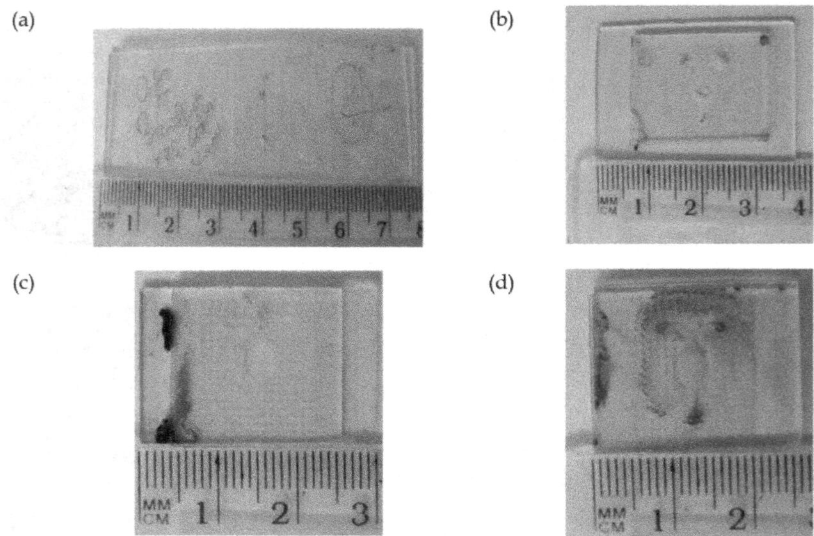

Figure 10. Some tested device using the microscope slide bonded in the oven. (a) a device is successfully bonded but has residues in it and (b) a successful bonded device with in 2 hours. (c) Successful bonded device in the oven 4 hours and (d) bonded device tested using dyes for leakage discovery.

3.9. The device

Fig. 11 shows one of the bonded microelectrodes array device that is ready for the experiment. The connecting wires is used to connect the device to the controller. The gold, Au and chromium, Cr layers are easily leached (dissolved) into the solder therefore soldering iron is not suitable to be used on gold, Au [30]. These wire connections are glued to gold, Au bonding pads, the large connection areas which were included in the microelectrode designs. The glue used is the silver paint or conductive epoxy glue from Chemtronics. It is cured at room temperature for about 4 hours. Table 3 shows the hours in the making of the epoxy conductive glue to cure under the room temperature (RT). This glue is suitable to be used when microelectrodes array is made from gold, Au or platinum, Pt which coated with titanium, Ti.

Hours on Bonding pad	Epoxy Conductive Glue
1 hour	Not cured
2 hour	Not cured
3 hour	Not cured
4 hour	Cured
5 hour	Cured

Table 3. Hours for the epoxy conductive glue to be cured in ambient air/room temperature (RT)

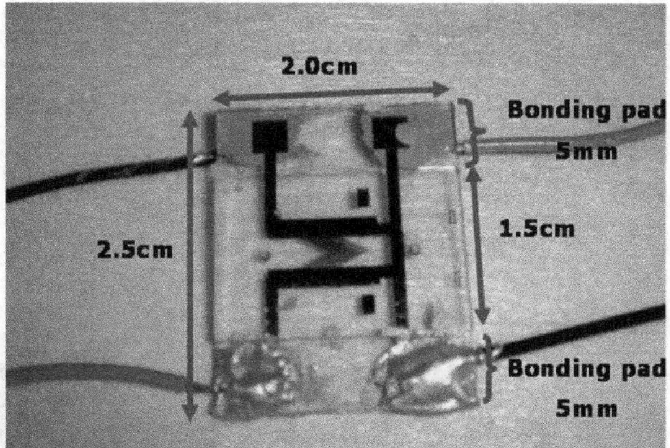

Figure 11. A bonded microelectrodes array with dimensions. The width of the device is 2.0 cm and length is 2.5 cm. The 5 mm length each side of the microfluidic device is made especially for bonding pad of both sided top and bottom microelectrodes [7].

3.10. Device holder

The holder is designed custom made for the device of size 20 mm x 20 mm shown in Fig. 12. The holder base is made from the PEEK material. The lid is made from brass material and it should be firmed enough to hold the device tightly and strongly. The gasket, which is made from polytetrafluoroethylene (PTFE), is a sheet or rubber type ring made of Teflon or a type of sealing material that can fill the space between two objects, the device and the holder, preventing leakage and breaking between the two objects, under compression/pressure.

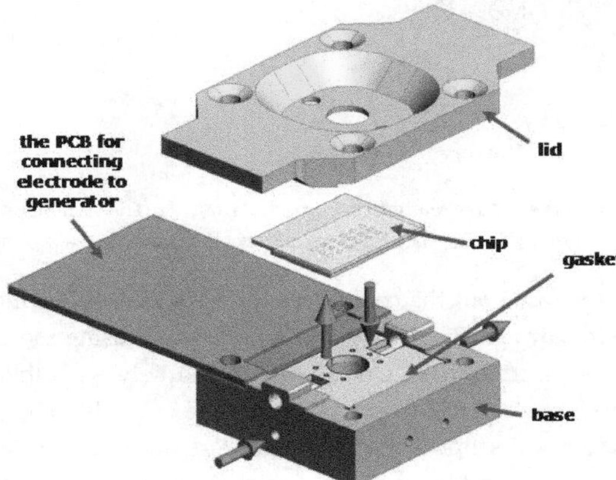

Figure 12. An illustration of a device/chip holder in 3D showing the lid, possible chip, gasket, base and the PCB connector to the electrode on the chip [illustrated by Rupert Thomas].

The device is clamped in the holder tightly as shown in Fig. 13. The holder can clamp the device of at least made of two substrates of each 700 μm thick and above.

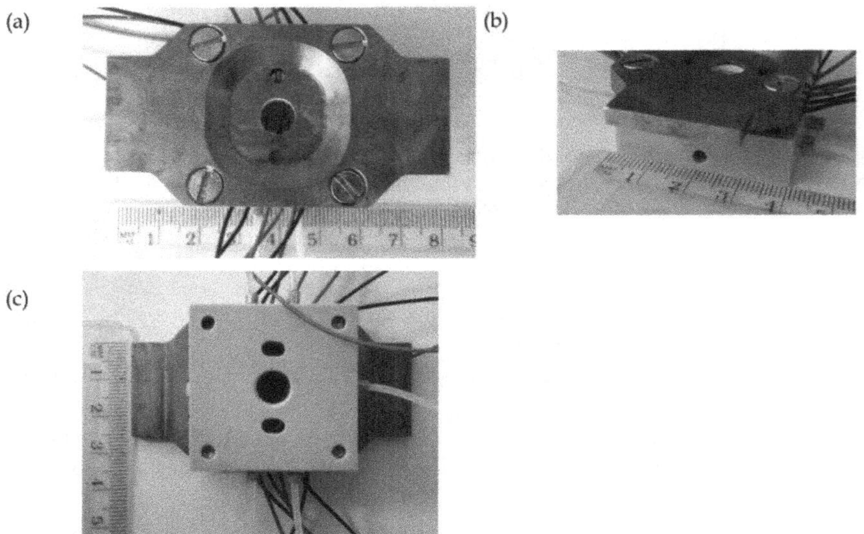

Figure 13. (a) The device in the holder ready to be experimented. (b) The dimension of the base of square shape is 4 mm x 4 mm (c) The top holder is turned upside down, showing the base of the holder.

The holder could be the same but the connectors to the device can be modified depending on the device connectors design. The connectors can be made using the cable of 26 pin and the working pin on the electrode could be selected. Uniquely with this holder, it can be used for viewing the channel of the device on either side of the holder as shown in Fig. 14 (a)-(b) as long as the microscope objective used can meet its working distance. The thickness of the base of the holder is 1 cm and the microscope objective used for this holder is X 10. Fig. 14 (c) and (d) show the dimension of the connectors, made from printed circuit board (PCB).

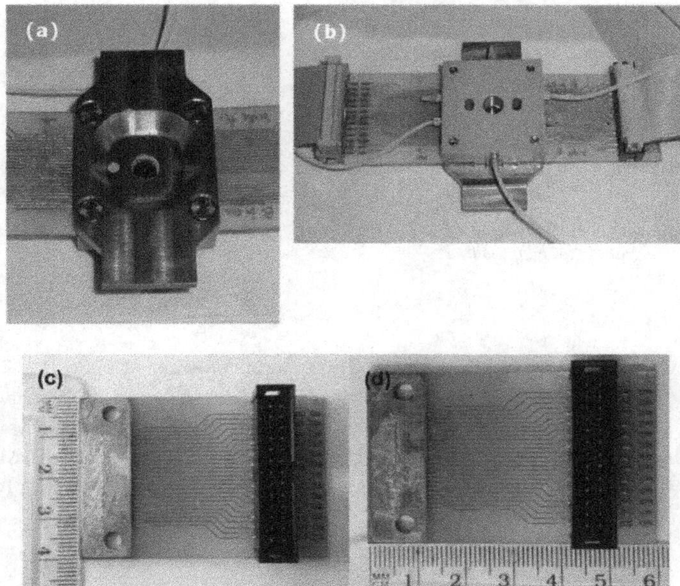

Figure 14. (a) The holder (b) The back of the holder with the connectors (c) and (d) show the dimension of the PCB connectors.

4. Discussion

The microfluidic device (from fragmented chip glasses) is successfully fabricated with an array of microelectrode in it and the microchannel made from the laminate sheet, dry film resists (DFR). The bonding process for the microchannel is seemed excellent to be done in the oven. The thermosetting material for this microchannel makes it can be cured in high temperature and also it cannot be dissolved easily after cured. This is a good indicator for a microfluidic device to be used for longer times in the experiment and during analysis. Hence the complete fabricated microfluidic devices depending on the microelectrode and microchannel designs, they can be used as a particle separator, mixer, focusing device and so on which definitely be functional in the bioelectrokinetic system.

5. Conclusion

In this chapter, the microfluidic device fabrication for bioelectrokinetic system applications is described. The microelectrode is fabricated using a standard microelectronic fabrication technique including the microchannel made of laminate sheet.

In order to make the system efficient, the steps of fabrication and cautions have had to be taken into account. All steps would require thorough precautions and might only need slight changes in time duration and it will depend on the size of the substrate, the thickness of the microelectrode and microchannel. In fabrication of microfluidic the microelectrode and microchannel dimensions are important for getting a good electrokinetic force to do manipulation of fluid and particle in bioelectrokinetic system [26].

Author details

Nurul Amziah Md Yunus
Universiti Putra Malaysia, Malaysia

Acknowledgement

The author would like to thank Prof. Hywel Morgan and Reader Dr. Nicolas G. Green for ideas, support and laboratory facilities during her research study at Bionanoelectronic Laboratory, Nano Group, School of Electronic and Computer Science (ECS), University of Southampton. A huge appreciation to Dr. Shahanara Banu, Dr. Rupert Thomas, Mr. Andrew Whitton and Ms. Katie Chamberlain for providing valuable discussion and advice to the author in making the microfluidic devices for bioelectrokinetic system applications until successful.

6. References

[1] Manz, A., N. Graber, and H.M. Widmer, *Miniaturized Total Chemical Analysis Systems: A Novel Concept for Chemical Sensing.* Sens. Actuators B, 1990. 1(1-6): p. 244-248.

[2] Reyes, D.R., et al., *Micro Total Analysis Systems. 1. Introduction, Theory, and Technology.* Analytical Chemistry, 2002. 74(12): p. 2623–2636.

[3] Lao, A.I.K., D. Trau, and I.M. Hsing, *Miniaturized Flow Fractionation Device Assisted by a Pulsed Electric Field for Nanoparticle Separation.* Analytical Chemistry, 2002. 74(20): p. 5364-5369.

[4] Terry, S.C., J.H. Jerman, and J.B. Angell, *A Gas Chromatographic Air Analyzer Fabricated on A Silicon Wafer.* IEEE Transactions on Electron Devices 1979. 26(12): p. 1880-1886.

[5] Shirinyan, A.S. and M. Wautelet, *Phase Separation in Nanoparticles.* Nanotechnology, 2004. 15(12): p. 1720-1731.

[6] Gascoyne, P.R.C. and J.V. Vykoukal, *Dielectrophoresis-based Sample Handling in General-Purpose Programmable Diagnostic Instruments.* Proceedings of the IEEE, 2004. 92(1): p. 22-42.

[7] Yunus, N.A.M. and N.G. Green, *Fabrication of Microfluidic Device Channel using A Photopolymer for Colloidal Particle Separation.* Microsystem Technologies, 2010: p. 1-9.

[8] Darrigol, O., *Worlds of Flow: A History of Hydrodynamics From the Bernoullis to Prandtl.* 2005: Oxford University Press, USA.

[9] Batchelor, G.K., *An Introduction to Fluid Mechanics.* 1967: Cambridge University Press Cambridge.

[10] Newton, I., A. Motte, and J. Machin, *The Mathematical Principles of Natural Philosophy*. 1729: Benjamin Motte.

[11] Stokes, G.G., *On the Effect of the Internal Friction of Fluids on the Motion of Pendulums*. 1851: Pitt Press.

[12] Acheson, D.J., *Elementary Fluid Dynamics*. 1990: Oxford University Press, USA.

[13] Morgan, H. and N. Green, *AC Electrokinetics: Colloids and Nanoparticles*. 2003: Research Studies Press Ltd.

[14] Reynolds, O., *An Experimental Investigation of the Circumstances Which Determine Whether the Motion of Water Shall Be Direct or Sinuous, and of the Law of Resistance in Parallel Channels*. Philosophical Transactions of the Royal Society of London, 1883. 174: p. 935-982.

[15] Rott, N., *Note on the History of The Reynolds Number*. Annual Review of Fluid Mechanics, 1990. 22(1): p. 1-12.

[16] Brody, J.P., et al., *Biotechnology at low Reynolds numbers*. Biophysical Journal, 1996. 71(6): p. 3430-3441.

[17] Derveaux, S., et al., *Synergism between Particle-based Multiplexing and Microfluidics Technologies may bring Diagnostics Closer to the Patient*. Analytical and Bioanalytical Chemistry, 2008. 391(7): p. 2453-2467.

[18] Sutera, S.P. and R. Skalak, *The History of Poiseuille's Law*. Annual Review of Fluid Mechanics, 1993. 25(1): p. 1-20.

[19] Manz, A., et al., *Planar Chips Technology for Miniaturization and Integration of Separation Techniques into Monitoring Systems: Capillary Electrophoresis on a Chip*. Journal of Chromatography, 1992. 593(1-2): p. 253-258.

[20] Harrison, D.J., et al., *Micromachining a Miniaturized Capillary Electrophoresis-based Chemical Analysis System on a Chip*. Science, 1993. 261(5123): p. 895-897.

[21] von Heeren, F., et al., *Micellar Electrokinetic Chromatography Separations and Analyses of Biological Samples on a Cyclic Planar Microstructure*. Analytical Chemistry, 1996. 68(13): p. 2044-2053.

[22] Huang, Y., et al., *Introducing Dielectrophoresis As A New Force Field for Field-Flow Fractionation*. Biophysical Journal, 1997. 73(2): p. 1118-1129.

[23] Pan, Y. and R. Yang, *A Glass Microfluidic Chip Adhesive Bonding Method at Room Temperature*. Journal of Micromechanics and Microengineering, 2006. 16(12): p. 2666-2672.

[24] Han, A.R., et al., *Multi-layer Plastic/Glass Microfluidic Systems Containing Electrical and Mechanical Functionality*. Lab on a Chip, 2003. 3(3): p. 150-157.

[25] Satyanarayana, S., R.N. Karnik, and A. Majumdar, *Stamp-and-Stick Room-Temperature Bonding Technique for Microdevices*. Journal of Microelectromechanical Systems, 2005. 14(2): p. 392-399.

[26] Yunus, N.A.M., H. Jaafar, and J. Jasni, *The Gradient of the Magnitude Electric Field Squared on Angled Microelectrode Array for Dielectrophoresis Applications (Special Issue: Asia-Pacific Symposium on Applied Electromagnetics and Mechanics (APSAEM10))*. Journal of the Japan Society of Applied Electromagnetics and Mechanics (JSAEM), 2011. 19: p. S240-S243.

[27] Nieuwenhuis, J.H. and M.J. Vellekoop. *Improved Dielectrophoretic Particle Actuators for Microfluidics.* in *Proceedings of IEEE Sensors* 2003.
[28] Yunus, N.A.M. and N.G. Green, *Continuous Separation of Submicron Particles Using Angled Electrode.* Journal of Physics: Conference Series Electrostatics 2007, 2007. 12: p. 012068.
[29] Metz, S., R. Holzer, and P. Renaud, *Polyimide-based Microfluidic Devices.* Lab on a Chip, 2001. 1(1): p. 29-34.
[30] Burt, R.J., *Semiconductor Chip Mounting System.* 1986, Google Patents

Permissions

The contributors of this book come from diverse backgrounds, making this book a truly international effort. This book will bring forth new frontiers with its revolutionizing research information and detailed analysis of the nascent developments around the world.

We would like to thank Mohammed A. A. Khalid, for lending his expertise to make the book truly unique. He has played a crucial role in the development of this book. Without his invaluable contribution this book wouldn't have been possible. He has made vital efforts to compile up to date information on the varied aspects of this subject to make this book a valuable addition to the collection of many professionals and students.

This book was conceptualized with the vision of imparting up-to-date information and advanced data in this field. To ensure the same, a matchless editorial board was set up. Every individual on the board went through rigorous rounds of assessment to prove their worth. After which they invested a large part of their time researching and compiling the most relevant data for our readers. Conferences and sessions were held from time to time between the editorial board and the contributing authors to present the data in the most comprehensible form. The editorial team has worked tirelessly to provide valuable and valid information to help people across the globe.

Every chapter published in this book has been scrutinized by our experts. Their significance has been extensively debated. The topics covered herein carry significant findings which will fuel the growth of the discipline. They may even be implemented as practical applications or may be referred to as a beginning point for another development. Chapters in this book were first published by InTech; hereby published with permission under the Creative Commons Attribution License or equivalent.

The editorial board has been involved in producing this book since its inception. They have spent rigorous hours researching and exploring the diverse topics which have resulted in the successful publishing of this book. They have passed on their knowledge of decades through this book. To expedite this challenging task, the publisher supported the team at every step. A small team of assistant editors was also appointed to further simplify the editing procedure and attain best results for the readers.

Our editorial team has been hand-picked from every corner of the world. Their multi-ethnicity adds dynamic inputs to the discussions which result in innovative outcomes. These outcomes are then further discussed with the researchers and contributors who give their valuable feedback and opinion regarding the same. The feedback is then collaborated with the researches and they are edited in a comprehensive manner to aid the understanding of the subject.

Apart from the editorial board, the designing team has also invested a significant amount of their time in understanding the subject and creating the most relevant covers. They scrutinized every image to scout for the most suitable representation of the subject and create an appropriate cover for the book.

The publishing team has been involved in this book since its early stages. They were actively engaged in every process, be it collecting the data, connecting with the contributors or procuring relevant information. The team has been an ardent support to the editorial, designing and production team. Their endless efforts to recruit the best for this project, has resulted in the accomplishment of this book. They are a veteran in the field of academics and their pool of knowledge is as vast as their experience in printing. Their expertise and guidance has proved useful at every step. Their uncompromising quality standards have made this book an exceptional effort. Their encouragement from time to time has been an inspiration for everyone.

The publisher and the editorial board hope that this book will prove to be a valuable piece of knowledge for researchers, students, practitioners and scholars across the globe.

List of Contributors

Mohammed Awad Ali Khalid
Department of Chemistry, College of applied medical and Science, University of Taif, Saudi Arabia Department of Chemistry, Faculty of Science, University of Khartoum, Sudan

V.E. Ptitsin
Institute for Analytical Instrumentation of the Russian Academy of Sciences, Saint - Petersburg, Russia

Yuichi Shimazaki
College of Science, Ibaraki University, Bunkyou, Mito, Japan

Ricardo Salgado
ESTS-IPS, Escola Superior de Tecnologia de Setúbal do Instituto Politécnico de Setúbal, Campus do IPS, Estefanilha, Setúbal, Portugal
REQUIMTE/CQFB, Chemistry Department, Faculdade de Ciências e Tecnologia da Universidade Nova de Lisboa, Campus de Caparica, Caparica, Portugal

Manuela Simões
CICEGE, Sciences of Earth Department, Faculdade de Ciências e Tecnologia da Universidade Nova de Lisboa, Campus de Caparica, Caparica, Portugal

Yoshihiro Kudo
Chiba University, Japan

Aoife C. Power and Aoife Morrin
National Centre of Sensor Research, School of Chemical Sciences, Dublin City University, Glasnevin, Dublin, Ireland

F. Robert-Inacio and G. Delafosse
Institut Supérieur d'Electronique et du Numérique, IM2NP
CNRS, IM2NP (UMR 7334)

L. Patrone
Institut Supérieur d'Electronique et du Numérique, IM2NP CNRS, IM2NP (UMR 7334)
Aix-Marseille Université, IM2NP, Maison des Technologies, Place Georges Pompidou, F-83000 Toulon, France

Nurul Amziah Md Yunus
Universiti Putra Malaysia, Malaysia

Printed in the USA
CPSIA information can be obtained
at www.ICGtesting.com
JSHW011402221024
72173JS00003B/396

9 781632 382375